STECK-VAUGHN

Pre-GED Science

Reviewers

Rochelle Kenyon
Assistant Principal
Adult and Vocational Off-Campus Centers
School Board of Broward County
Fort Lauderdale, Florida

Dee Akers Prins
Resource Specialist in Adult Education
Richmond Public Schools
Richmond, Virginia

Danette S. Queen
Adult Basic Education
New York City Public Schools
New York, New York

Margaret A. Rogers
Winterstein Adult Center
San Juan Unified School District
Sacramento, California

Lois J. Sherard
Instructional Facilitator
Office of Adult and Continuing Education
New York City Board of Education
New York, New York

STECK-VAUGHN COMPANY
A Subsidiary of National Education Corporation

Acknowledgments

Executive Editor: Elizabeth Strauss
Supervising Editor: Carolyn Hall
Editor: Susan Miller
Design Director: D. Childress
Design Coordinator: Cynthia Ellis
Cover Design: D. Childress
Editorial Development, Design/Production: McClanahan & Company, Inc.
Project Director: Mark Moscowitz
Writers: Marion B. Castelluci, Eugene R. Fox, Kathryn R. Fox
Editor: Kathryn R. Fox
Design/Production: McClanahan & Company, Inc.

Photograph Credits: Cover: © Masahiro Sano/The Stock Market
p. 12 © Alan Pitcairn/Grant Heilman
p. 13 © Ken Regan/Camera Five
pp. 15–16 NASA
p. 21 American Cancer Society
p. 27 A. Jalandon/Monkmeyer Press
p. 28 Carolina Biological
p. 33 © Bob Combs/Photo Researchers
p. 39 Michael Kevin Daly/Stock Market
p. 45 Michael Kienitz / © 1989 Discover Magazine
p. 51 © Bachmann/The Imageworks
p. 53 March of Dimes
p. 59 © David Stone/Photo Nats
p. 65 © Horst Schafer/Peter Arnold

Illustration Credits: Maryland Cartographics, Inc. pages 1, 2, 4, 6, 8, 64, 71, 82, 83, 102, 103, 120, 130, 139, 151, 156, 162, 163, 169, 174, 187, 191, 192, 193, 198, 201, 203, 206
University Graphics pages 22, 29, 34, 40, 41, 46, 47, 52, 58, 70, 76, 77, 89, 92, 93, 96, 107, 113, 125, 137

Credits continue on page 236, which is an extension of this copyright page.

ISBN 0-8114-4487-2

Printed in the United States of America.

2 3 4 5 6 7 8 9 PO 98 97 96 95 94 93 92

Table of Contents

To the Student

How To Use This Book

This book allows you to build upon what you already know to improve your science skills. You will increase your knowledge and understanding of the four areas of science by reading interesting articles on many different topics. These topics are divided into the four units described below.

Units

Unit 1: Life Science. Life science is the study of living things, where they live, and how they affect each other. In this unit, you will develop such science skills as making inferences and predictions, classifying, and comparing and contrasting. The graphic illustrations in this unit also provide practice in skills such as understanding photos and diagrams, and reading maps, timelines, and graphs. This unit contains articles about the human body ranging from skin cancer to genetic screening.

Unit 2: Earth Science. Earth science is the study of Earth and its surroundings. This unit covers science skills such as distinguishing fact from opinion and reading tables and weather maps. You will read about weather, the greenhouse effect, water conservation, rocks, and outer space.

Unit 3: Chemistry. Chemistry is the study of matter and how it changes. This unit includes science skills such as understanding chemical formulas, comparing and contrasting, making predictions, and reading line graphs. You will gain an understanding of chemistry by reading articles about the chemical nature of household cleaners, chemical reactions in cooking, and nuclear reactions.

Unit 4: Physics. Physics is the study of energy and how it affects matter. In this unit, you will practice reading diagrams, making inferences, summarizing information, and drawing conclusions. You will read articles about how machines work, force and motion, electrical safety in your home, and lasers.

Inventory and Posttest

The Inventory is a self-check of what you already know and what you need to study. After you complete all of the items on the Inventory, the Correlation Chart tells you where each skill is taught in this book. When you have completed the book, you will take a Posttest. Compare your Posttest score to your Inventory score to see your progress.

Sections

All the units are divided into sections. Each section is based on the Active Reading Process. *Active reading* means doing something *before reading, during reading,* and *after reading.* By reading actively, you will improve your reading comprehension skills.

Setting the Stage

Each section begins with an activity that helps you prepare to read the article. This is the activity you do *before reading.* First, determine what you already know about the subject of the article. Then, preview the article by reading and writing the headings in the article. Finally, write the questions that you expect the article will answer.

The Article

The articles you will read are about interesting topics in science. As you read each article, you will see a feature called *Applying Your Skills and Strategies.* Here you learn a reading or science skill, and you do a short activity. After completing the activity, continue reading the article. *Applying Your Skills and Strategies* occurs twice in every article. These are the activities you do *during reading.*

Thinking About the Article

These are the activities you do *after reading.* Here you answer fill-in-the-blank, short-answer, or multiple-choice questions. Answering these questions will help you decide how well you understood what you just read. The final question in this section relates information from the article to your own real-life experiences.

Answers and Explanations

Answers and explanations to every exercise item are at the back of this book, beginning on page 211. The explanation for multiple-choice exercises tells why one answer choice is correct and why the other answer choices are incorrect.

Study Skills

Good study skills are important. Here are some things you can do to improve your study skills.

- Find a quiet place to study.
- Organize your time by making a schedule.
- Take notes by restating important information in your own words.
- Look up any words you don't know in a dictionary or the glossary at the end of this book.
- Make a list of concepts and skills on which you need to work. Take time to go back and review this material.

INVENTORY

Use this Inventory before you begin Section 1. Don't worry if you can't answer all the questions. The Inventory will help you find out which skills you are strong in and which skills you need to practice. Read each article, study any graphics, and answer the questions that follow. Check your answers on page 211. Then enter your scores on the chart on page 11.

The Plant Cell

All living things are made of cells, the working units of the body. Plant cells differ from animal cells. Plant cells have a cell wall for strength. They also have a chemical called chlorophyll for making food. The food gives the plant cell energy and substances needed for growth.

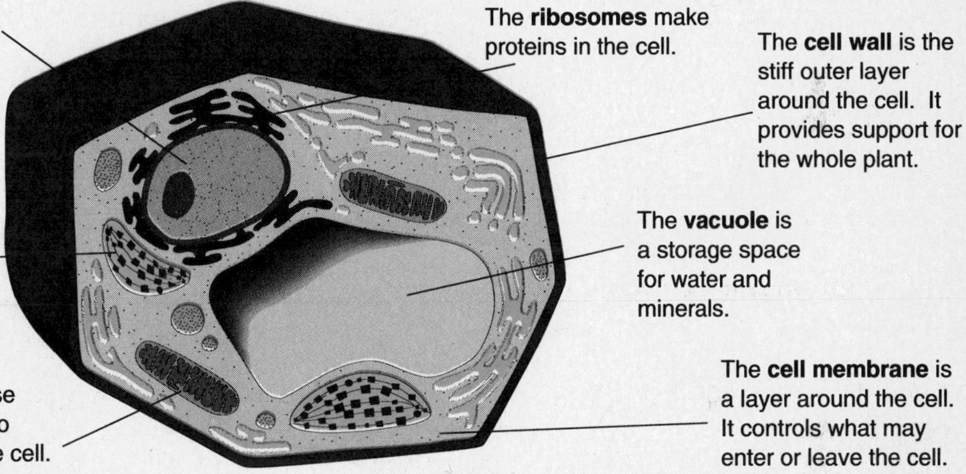

Fill in the blank with the word or words that best complete each statement.

1. The control center of a cell is its ______________.

2. The ________________ gives the plant cell strength.

Circle the number of the best answer.

3. A plant cell needs water when its
   (1) nucleus is empty.
   (2) chloroplast is empty.
   (3) vacuole is empty.
   (4) ribosome is empty.
   (5) membrane is empty.

Go on to the next page.

The Muscles of the Arm

Pick up a cup of coffee. You bend your elbow and raise your lower arm. This action is caused by a muscle in the upper arm. Now put down the cup of coffee. You straighten your elbow and move your lower arm down. This action is caused by another muscle in the upper arm.

Muscles work in pairs. The biceps muscle bends the elbow joint. The triceps muscle straightens the elbow joint. Why does it take two muscles to operate one joint? Muscles pull, but they cannot push. A muscle works by contracting, or shortening. When the biceps contracts, it pulls on the bones of the lower arm. The elbow joint bends. When the triceps contracts, it pulls on the same bones but in the opposite direction. When one muscle is contracting, its partner is relaxing.

Biceps relaxes

Triceps contracts

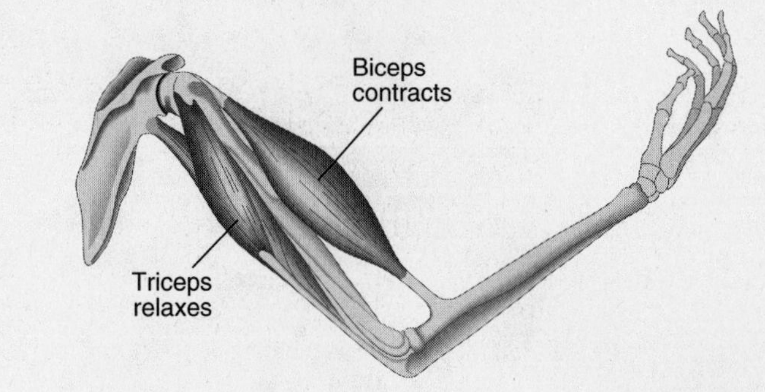

Fill in the blank with the word or words that best complete each statement.

4. When a muscle is ____________________, it gets shorter.
5. When one muscle in a pair is contracting, the other muscle is

   ________________.

Circle the number of the best answer.

6. The large muscles at the back of the thigh bend the knee joint. Where are the muscles that straighten the knee joint?

   (1) at the back of the thigh

   (2) at the front of the thigh

   (3) at the front of the lower leg

   (4) at the back of the lower leg

   (5) at the hip

Go on to the next page.

Bacteria

Bacteria are simple one-celled organisms. They can be found just about anywhere. The action of bacteria can be good for people, or it can be harmful. The bacteria that turn milk into cheese or yogurt are useful. These same bacteria also turn the milk in your refrigerator sour. Then they are not so useful.

If sour milk and yogurt are made by the same bacteria, why does one taste so bad? The difference is how the bacteria are controlled. When yogurt is made, the bacteria are killed before they make the product too sour. In your refrigerator, the bacteria keep on going. The result is that the milk gets much too sour. The bacteria that make milk sour use the sugar in the milk for energy. Their waste product, lactic acid, is the sour substance you taste in the spoiled milk.

The activity of bacteria depends on the temperature. People are often surprised when they find that food has gone bad in the refrigerator. Keeping food cool does slow down the bacteria. However, it does not stop their action. Freezing food does stop the action of bacteria. But it does not kill the bacteria. So food that has been in the freezer can spoil after it has been defrosted.

Bacteria can be killed by high temperatures. Milk and other dairy products are pasteurized. In this process, the milk is heated to a high temperature, then quickly cooled. The heat kills the bacteria. But once you take the milk home and open it, new bacteria may get in. Then the spoiling process begins.

Fill in the blank with the word or words that best complete each statement.

7. ____________________, which is a waste product of bacteria, makes spoiled milk taste sour.

8. Milk is turned into yogurt by the action of ________________.

Circle the number of the best answer.

9. How does the pasteurizing process affect bacteria?

    (1) Pasteurizing kills bacteria with high temperatures.

    (2) Pasteurizing kills bacteria with low temperatures.

    (3) Pasteurizing slows bacteria with low temperatures.

    (4) Pasteurizing slows bacteria with high temperatures.

    (5) Pasteurizing poisons bacteria with chemical preservatives.

Go on to the next page.

The Flower

The reproductive organ of a plant is the flower. Flowers come in many shapes and sizes. Some are large and bright. Others are so small, you might not notice them.

Flowers make pollen. The pollen contains sperm, which is made in the anther of the flower. The pollen is carried from one flower to another by insects, birds, or the wind. When pollen from one flower reaches another flower of the same kind, the sperm in the pollen joins with the egg which is made in the ovary of the flower. The fertilized egg becomes a seed. The seed begins the next generation.

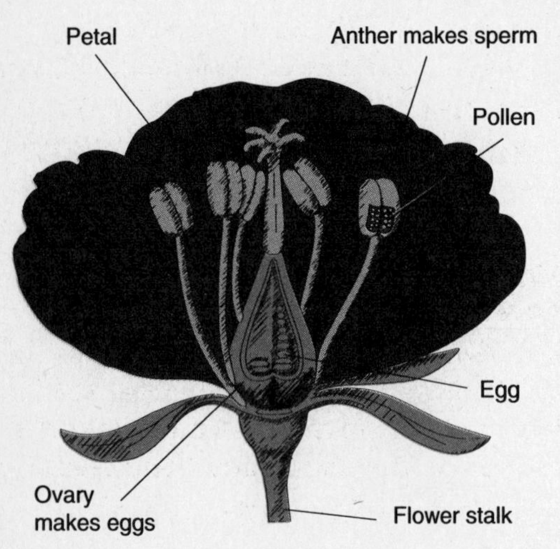

Flowers that are pollinated by insects, such as bees, often are bright in color. They have large petals, which give the bees a place to land. These flowers also make nectar. This sweet juice attracts bees to the flower. When the bees go to drink the nectar, some of the sticky pollen gets on their bodies. This pollen rubs off when the bees get to the next flower.

The flowers of many trees and grasses are pollinated by the wind. These flowers usually have tiny petals. Some have no petals at all. The pollen is dry and dusty. These conditions make it easy for the pollen to blow from one plant to another.

Identify each of the following as a characteristic of a plant that is pollinated by insects or by the wind.

10. Sticky pollen ______________________

11. Sweet nectar ______________________

12. Small petals ______________________

13. The flower shown on this page ______________________

Circle the number of the best answer.

14. Which part or parts of the flower are necessary for reproduction?
    (1) petals
    (2) anthers
    (3) ovary
    (4) petals and anthers
    (5) anthers and ovary

Go on to the next page.

Biomes

You probably know that Earth's climate is warmest near the equator and coldest near the poles. The climate determines the living things that are found in any area. Large regions with the same climate throughout have the same kinds of living things. A large region with a certain climate and certain living things is called a biome.

There are many kinds of biomes on Earth. Most of the eastern and northeastern United States is a forest biome. This biome is home to many kinds of trees and birds. There are also many animals, such as deer, foxes, squirrels, and chipmunks.

Most of the central United States is a grassland biome. Because this biome is drier than the forest biome, there are few trees. The animals of the grassland feed on the many kinds of grasses there. Once there were huge herds of bison grazing the grasslands. Now there are herds of sheep and cattle. The most common wild animals are small. There are rabbits, prairie dogs, and badgers.

Much of the southwestern United States is a desert biome. It is so dry that few plants can grow there. Rain may be heavy in the desert, but it does not rain often or for long. Then it dries up quickly. The cactus plant survives here because it is able to store water. Most of the animals of the desert live underground to escape the heat. They come out at night or early morning when it is cool. Small animals feed on the plants and their seeds. These animals are food for coyotes, hawks, and rattlesnakes.

Fill in the blank with the word or words that best complete each statement.

15. The ____________ biome is home to many trees and deer.

16. The __________________ biome is home to rabbits and prairie dogs.

17. A desert plant that can store water is the ____________.

Circle the number of the best answer.

18. The biomes of the United States change as you move from east to west because the climate becomes

    (1) cooler.

    (2) drier.

    (3) windier.

    (4) less sunny.

    (5) all of the above.

Go on to the next page.

The Water Cycle

When you are caught in a sudden rainstorm, you probably don't think about where all that water came from. But you are experiencing one step in the water cycle. The water cycle is the circulation of water on Earth and in its atmosphere.

Water covers more than half of the planet. This surface water is found in oceans, lakes, and rivers. Water constantly evaporates from surface water. Water in the atmosphere condenses and forms clouds. When the clouds become too heavy, the water falls as rain. Rain, snow, and sleet are forms of precipitation. When it rains, some water soaks into the ground and some moves along the land to rivers and lakes.

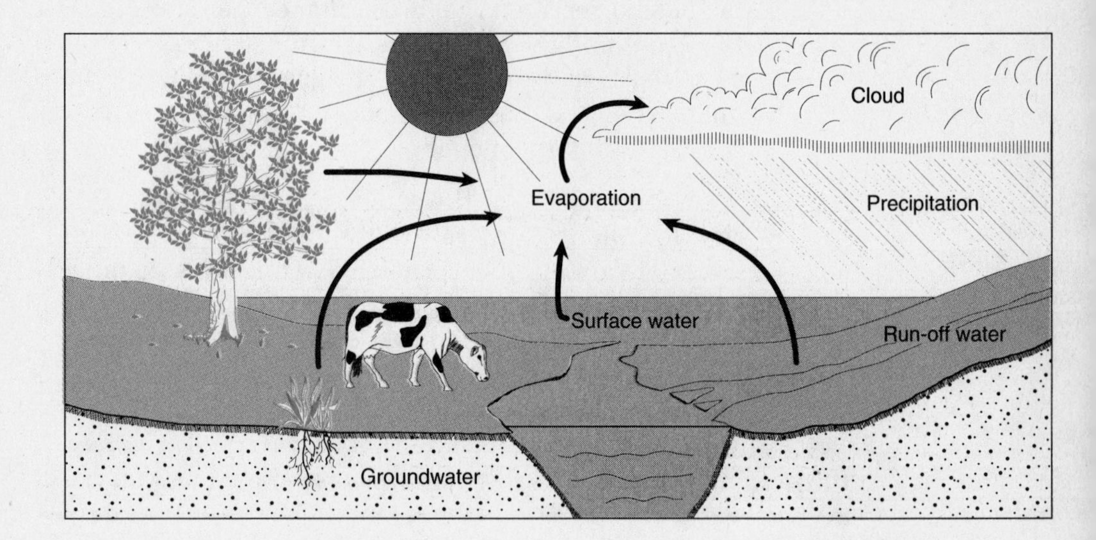

Match the part of the water cycle with its description. Write the letter of the part of the cycle in the blank at the left.

		Description		**Part of the Water Cycle**
______	19.	rain and snow	a.	clouds
______	20.	lakes and oceans	b.	groundwater
______	21.	water that soaks into the soil	c.	precipitation
______	22.	water that condenses in the atmosphere	d.	run-off water
______	23.	water flowing over the top of the soil	e.	surface water

Go on to the next page.

Rocks

Have you ever washed your hands with gritty soap? The soap probably had ground pumice in it. Pumice is a kind of rock. This rock forms when lava from a volcano cools and hardens. Igneous rocks, such as pumice and granite, form when melted rock hardens. Some buildings are made of granite.

Over a long period of time, wind and water can wear down rocks. Small pieces of the rocks are blown or washed away. These pieces may settle slowly and form layers. The particles are called sediment. Slowly, the layers harden, forming sedimentary rocks. Sandstone, limestone, and shale are sedimentary rocks. Sedimentary rocks are not as hard as igneous rocks. Sandstone and limestone wear away much faster than granite.

Igneous and sedimentary rocks can be changed into new forms. This is caused by high temperatures or great pressures. Rocks formed in this way are called metamorphic rocks. Marble is a metamorphic rock. You may have seen statues made of marble.

Fill in the blank with the word or words that best complete each statement.

24. Rocks that form as layers of hardened particles are called

    ________________________________.

25. Rocks that form under high temperatures or pressures are called

    __________________________.

Circle the number of the best answer.

26. Which of these is an example of an igneous rock?

    (1) granite

    (2) limestone

    (3) marble

    (4) sandstone

    (5) none of the above

27. An area where you find many igneous rocks once may have had

    (1) many rivers.

    (2) large amounts of sediment.

    (3) high pressure.

    (4) volcanoes.

    (5) earthquakes.

Go on to the next page.

The Atom

All matter is made up of atoms. An atom is made up of three types of particles. Protons are particles with a positive charge. Neutrons have no charge. Protons and neutrons are found in the nucleus, or core, of an atom. All atoms of a particular element have the same number of protons. The number of neutrons can vary.

Orbiting around the nucleus are negatively charged particles, called electrons. Electrons are much lighter than protons or neutrons. The number of electrons in an atom is equal to the number of protons, so an atom has no charge. If an atom gains or loses an electron, it is called an ion. An ion has a charge.

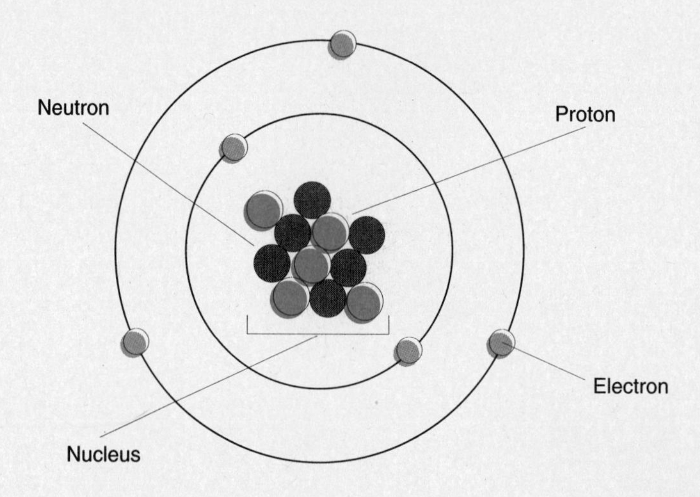

Match the name of the particle with its description. Write the letter of the particle in the blank at the left. Letters may be used more than once.

	Description	**Particle**
______	28. has no charge	a. proton
______	29. not found in the nucleus	b. neutron
______	30. has a positive charge	c. electron
______	31. a negatively charged ion has an extra one	
______	32. all atoms of an element have the same number of these	

Go on to the next page.

States of Matter

On Earth, all matter is found in three states, solid, liquid, and gas. The table below shows the properties of these states. The state of matter of a substance depends on its temperature. Each substance changes state at different temperatures. At low temperatures, substances are solids. Sugar, salt, and plastic are solids at room temperature. If a solid is heated, it changes to a liquid. Water and alcohol are liquids at room temperature. If a liquid is heated, it changes to a gas. Air is a mixture of substances that are gases at room temperature.

State	Temperature Range	Shape	Volume
solid	lowest	definite (does not change)	definite (does not change)
liquid	middle	not definite; takes the shape of its container	definite (does not change)
gas	highest	not definite; takes the shape of its container	not definite; expands to fill the volume of its container

Circle the number of the best answer.

33. In which state or states of matter does the volume stay the same?
    (1) solid only
    (2) liquid only
    (3) gas only
    (4) solid and liquid only
    (5) solid, liquid, and gas

34. What happens if you cool a liquid enough to make it into a solid?
    (1) Its shape changes from not definite to definite.
    (2) Its shape changes from definite to not definite.
    (3) Its volume changes from not definite to definite.
    (4) It will be in its highest temperature range.
    (5) None of its properties changes.

Go on to the next page.

Forces

A force is a push or pull. In a tug of war, two teams pull on a rope. Each team exerts a force on the rope. The forces act in opposite directions. If the two forces are equal, there is no change in movement. The forces are said to be balanced.

Suppose an extra player joins one team. The extra player increases that team's force. Now the forces are not balanced, and the movement of the rope changes. A change in movement is called acceleration. The rope accelerates, or moves, in the direction of the greater force. The team with the extra player wins.

If you hold a ball in your hand, the forces on the ball are balanced. Gravity pulls down on the ball. Your hand exerts an upward force that balances the force of gravity. If you let go of the ball, your force changes, but gravity does not. Gravity "wins," and the ball accelerates in the direction of the pull of gravity.

If you throw a ball, the forces involved get more complicated. When you throw the ball, you exert a force on it. The forward force of your hand is opposed by the force of friction between the ball and the air. These forces are not balanced. Because your force is greater, the ball moves in the direction you throw it. Once the ball leaves your hand, you exert no force on it. Friction continues to act on the ball. This force is not balanced, so it changes the motion of the ball. The ball slows down. At the same time, gravity pulls on the ball. This downward force is not balanced by another force. So the ball falls down as it moves forward.

Fill in the blank with the word or words that best complete each statement.

35. A __________ is a push or a pull.

36. A change in movement is called ______________________________.

Circle the number of the best answer.

37. Which of the following is an effect of an unbalanced force?

(1) acceleration

(2) gravity

(3) friction

(4) a push

(5) a pull

Check your answers on pages 211–212.

INVENTORY
Correlation Chart

Science

The chart below will help you determine your strengths and weaknesses in the four content areas of science.

Directions

Circle the number of each item that you answered correctly on the Inventory. Count the number of items you answered correctly in each row. Write the amount in the Total Correct space in each row. (For example, in the Life Science row, write the number correct in the blank before *out of 18*). Complete this process for the remaining rows. Then add the 4 totals to get your total correct for the whole 37-item Inventory.

Content Areas	Items	Total Correct	Pages
Life Science (Pages 12–97)	1, 2, 3 4, 5, 6 7, 8, 9 10, 11, 12, 13, 14 15, 16, 17, 18	_____ out of 18	Pages 20–25 Pages 38–43 Pages 44–49 Pages 56–61 Pages 80–85
Earth Science (Pages 98–133)	19, 20, 21, 22, 23 24, 25, 26, 27	_____ out of 9	Pages 112–117 Pages 118–123
Chemistry (Pages 134–169)	28, 29, 30, 31, 32 33, 34	_____ out of 7	Pages 136–141 Pages 142–147
Physics (Pages 170–199)	35, 36, 37	_____ out of 3	Pages 172–177
TOTAL CORRECT FOR INVENTORY _____ out of 37			

If you answered fewer than 34 items correctly, look more closely at the four content areas of science listed above. In which areas do you need more practice? Page numbers to refer to for practice are given in the right-hand column above.

LIFE SCIENCE

Ocotillo and saguaro cactus growing in the desert

Life science is the study of living things. It includes how living things work and where they live. Life science is also about how living things affect one another. Knowing about life science helps you understand yourself and the world around you.

One part of life science is the study of the human body. You may have seen articles about health and medicine in newspapers and magazines. Knowing more about how the body works will help you understand these articles. Knowing about the body will also help you to stay healthy.

Life science includes the study of plants and animals. Plants and animals provide food and many other useful things. But some plants and animals cause problems. You may have had a problem with insect pests. Knowing about life science will help you solve these problems.

Life science includes the study of the environment. People are facing many decisions about their world. What should be done with garbage? Why is it important to protect our forests and beaches? As you learn more about life science, you will become aware of these problems. You will also become aware of ways to solve them.

Life scientists are the people who study life science. These scientists have added much to our understanding of the world. Jane Goodall is a life scientist who studies animals. She has spent thirty years studying chimpanzees in Africa.

To get to Africa, Goodall became the secretary of a scientist there. She had studied animals before on her own. She was so set on being a scientist that her boss suggested she study chimpanzees.

At first, the chimps ran away when Goodall came near them. Eight months passed before they allowed her to sit nearby and watch them. By looking and keeping careful records, Goodall learned many things. She discovered that chimps often act like humans. Mother chimps cuddle their babies. When friends meet, they hug or hold hands. One of the most surprising things Goodall learned was that chimps use simple tools.

Goodall has written books about chimps and other animals. Along the way, Goodall returned to school to get her Ph.D. degree. Dr. Goodall still lives in Africa, where she continues to study and write about animals.

Jane Goodall

This unit features articles about many areas of life science and the work of many life scientists.

- The human body articles include information about pregnancy, exercise, and cutting down on cholesterol.
- Plant and animal articles include topics such as wolves, and medicines from plants.
- Articles about the environment include discussions of rain forests and the garbage crisis.

Section 1

The Scientific Method

Setting the Stage

Can people live, work, and have children in outer space? Perhaps they can. To begin to answer such a question, scientists gather information and test ideas.

Past: What you already know

You may already know something about space or about gathering information. Write two things you already know.

1. ______________________________

2. ______________________________

Present: What you learn by previewing

When **previewing** an article, look it over quickly to find out what it is about. Do not read it. Instead, look at headings, pictures, and diagrams to get an idea of what the article is about. Write the headings from the article on pages 15–17 below. The first one is done for you.

An Experiment in Space

The Scientific Method ______________ 5. ______________

3. ______________ 6. ______________

4. ______________

What does the photo on page 17 show?

7. ______________________________

____

Future: Questions to answer

Write two questions you expect this article to answer.

8. ______________________________

9. ______________________________

Check your answers on page 212.

An Experiment in Space

As you read each section, circle the words you don't know.

When people fly in space, they do not feel the pull of Earth's gravity. Being weightless may be fun, but it can also cause problems. Some astronauts get space-motion sickness. Others come back from long flights with bones and muscles that are weaker.

It is important to understand how weightlessness affects the human body. Space flights to distant planets may take years. People may live and work for long periods in space stations. Will they be able to raise their own animals for food? Will they be able to have children?

Weightless astronauts outside space shuttle *Atlantis*

The Scientific Method

Questions like these are not easy to answer. To find answers, scientists have a process for getting information and testing ideas. This process is called the scientific method. The **scientific method** helps scientists solve problems logically. The scientific method includes four main steps. These steps are observation, hypothesis, experiment, and conclusion.

Observation

A student named John Vellinger became curious about weightlessness. Through **observation**, he learned about what was already known. He watched astronauts on TV. He read as much as he could. He wondered how weightlessness would affect people in space for long periods. Would they be able to raise animals for food? Would they be able to have children?

Hypothesis

Once scientists ask questions, they try to guess the answers. A **hypothesis** is a guess about the answer to a question. A hypothesis is not a wild guess. It is based on many observations. Would weightlessness affect the development of an **embryo**, a living thing in an early stage of life? After making observations, you might guess that an embryo would not be affected by weightlessness. This guess would be your hypothesis.

Applying Your Skills and Strategies

Making Predictions. When scientists state a hypothesis, they are predicting—or guessing—what will happen. You can make predictions based on what you read. Write two predictions based on what you have read so far.

____

____

____

____

____

____

____

Experiment

Scientists must test whether their hypothesis is right. They do this with an **experiment**. Vellinger designed an experiment to test if weightlessness would affect chicken embryos.

Vellinger decided to use 64 fertilized chicken eggs. He divided the eggs into two groups. One group of eggs would spend five days in space on the shuttle *Discovery*. This group was the **experimental group**. The other group of eggs would stay on Earth. The second group was the **control group**. The two groups differed in only one way—the experimental eggs would be weightless for five days.

Vellinger then divided both the experimental and control groups. Sixteen eggs from each group were fertilized nine days before the shuttle launch. The other sixteen eggs were fertilized two days before the launch.

What happened in the experiment? If our hypothesis was right, all 64 eggs should hatch after the normal 21-day development period. Instead, something surprising happened. None of the younger embryos aboard the shuttle survived. Yet the older group that had also spent time in space hatched normally.

 Check your answers on page 212.

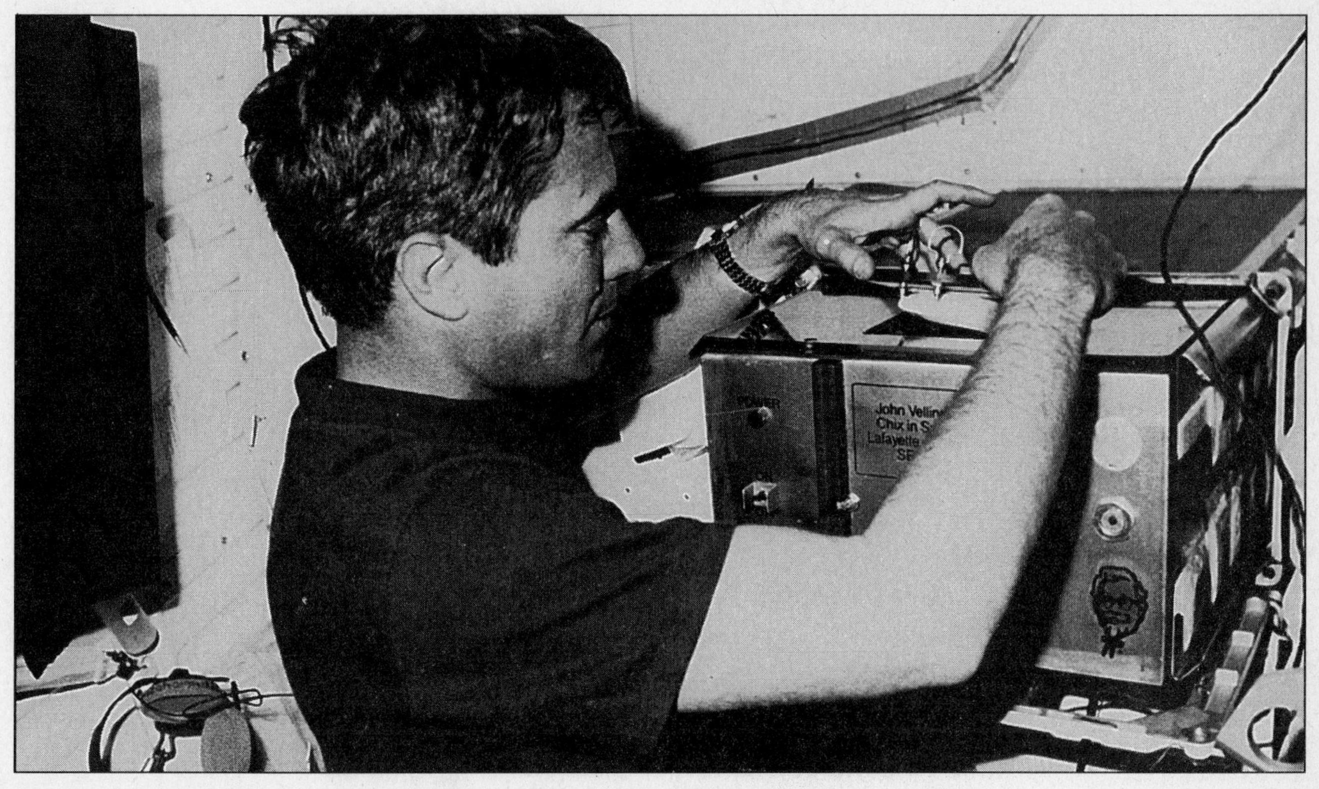

Astronaut John E. Blaha checks an incubator on the space shuttle *Discovery*.

Applying Your Skills and Strategies

Using the Glossary or Dictionary. When you read, you may see words you do not understand. Circle any words you do not know. You can look up the meaning of words in dark print in the glossary at the back of this book. Other unfamiliar words can be found in a dictionary.

Circle all of the words in dark print in this article. Write their definitions on your own paper. Then look up any other words you do not know. Write their definitions on your own paper, too.

Conclusion

The results of an experiment are stated in a **conclusion**. This experiment shows that our hypothesis was wrong. Our conclusion is that weightlessness seems to affect the growth of embryos. But it does not seem to affect embryos at different stages of growth in the same way. The embryos that did not hatch were all in the first third of the 21-day development period. The embryos that hatched were in the second third of the development period when they were weightless.

Astronaut John E. Blaha, the pilot of *Discovery*, said, "What looked like a simple experiment may have . . . generated thousands of questions." With the information from this experiment, scientists will make new hypotheses about weightlessness. More experiments will be done. Someday scientists will be able to answer the question: can people live, work, and have children in outer space?

Check your answers on page 212.

Thinking About the Article

Fill in the blank with the step of the scientific method that best completes each statement.

1. While Sam is cooking hot dogs, the phone rings and he leaves the room. In a few minutes, he smells something burning. He is making an

   ____________________.

2. Because of the smell, Sam immediately guesses that the hot dogs are burning. This is his ____________________.

3. To see whether he has guessed correctly, Sam rushes into the kitchen. He finds that the hot dogs are fine, but the potholder is on fire. This is his ____________________.

4. Sam's guess was wrong. After putting out the fire, he comes to the ____________________ that he left the potholder too close to the stove.

Write your answers in the space provided.

5. Review the questions you wrote on page 14. Did the article answer your questions? If you said *yes*, write the answers. If your questions were not answered, write three things you learned from this article.

____

____

____

____

____

6. Why is it important to understand the effects of weightlessness on people?

____

____

____

____

Check your answers on page 212.

7. What were the results of the chicken egg experiment?

Circle the number of the best answer.

8. The scientific method is a way to
   (1) raise money for scientific experiments.
   (2) solve problems in a logical way.
   (3) measure information in an experiment.
   (4) protect people from the effects of weightlessness.
   (5) predict what will happen.

Write your answers in the space provided.

9. Why is it important that the experimental and control groups in an experiment differ in only one way?

10. Describe an experience you or someone you know has had with doing an experiment or solving a problem. Write your description in complete sentences.

11. Would you like to travel in space? Give a reason for your answer.

Check your answers on page 213.

Section 2

The Cell

Setting the Stage

Each year many people in the United States get skin cancer. Most skin cancer is caused by too much sun. Harmful rays in sunlight can cause skin cells to grow in abnormal ways.

Past: What you already know

You may already know something about cancer or the effects of sunlight. Write two things you already know.

1. ____________________

2. ____________________

Present: What you learn by previewing

When **previewing** an article, look it over quickly to find out what it is about. Do not read it. Instead, look at headings, pictures, and diagrams to get an idea of what the article is about. Write the headings from the article on pages 21–23 below. The first one is done for you.

Sunlight and Skin Cancer

Types of Skin Cancer

3. ____________________

4. ____________________

What does the diagram on page 22 show?

5. ____________________

Future: Questions to answer

Write three questions you expect this article to answer.

6. ____________________

7. ____________________

8. ____________________

Check your answers on page 213.

Sunlight and Skin Cancer

As you read each section, circle the words you don't know. Look up the meanings.

Each year, millions of Americans relax at beaches, swimming pools, and in the mountains. Many people work outdoors in the sun. Some people even go to tanning parlors to get a bronzed look. However, exposure to **ultraviolet light**, a type of light in sunshine, can cause skin cancer. In fact, there are about 600,000 new cases of skin cancer in the United States each year. That makes skin cancer the most common cancer in this country. Skin cancer is also the easiest form of cancer to treat and cure.

Types of Skin Cancer

There are three types of skin cancer. They are called basal cell, squamous cell, and melanoma.

The most common and least dangerous skin cancer is **basal cell skin cancer**. About 450,000 people develop this type of skin cancer each year. Basal cell skin cancer often appears on the hands or face. It may look like an open sore, reddish patch, mole, shiny bump, or scar. Basal cell skin cancer grows slowly and rarely spreads to other parts of the body. When found early and removed, basal cell skin cancer can almost always be cured.

Squamous cell skin cancer is more dangerous than basal cell skin cancer. It affects about 115,000 people each year. Squamous cell skin cancer looks like raised, pink spots or growths that may be open in the center. This cancer grows faster than basal cell skin cancer. Squamous cell skin cancer can spread to other parts of the body. If it is not treated, this type of skin cancer may lead to death.

By far the most dangerous of the three types of skin cancer is **melanoma**. There are 32,000 new melanoma cases each year. This type of skin cancer may grow in a mole or on clear skin. Melanomas are oddly shaped blotches that turn red, white, or blue in spots. Melanomas become crusty and bleed, and they grow fast. When melanomas reach the size and thickness of a dime, it's likely that they have spread and become deadly. About 6,500 people die of melanoma each year in the United States.

Applying Your Skills and Strategies

Finding the Main Idea. One way to make sure you understand what your read is to find the main idea of each paragraph. A paragraph is a group of sentences about one main idea or topic. The main idea is usually stated in one sentence called a *topic sentence*. The topic sentence is general enough to cover all the points made in the paragraph.

The topic sentence of the paragraph above has been underlined. Underline the topic sentence in each of the last two paragraphs under the heading *Types of Skin Cancer.*

How Sunlight Causes Skin Cancer

The human body is made of many cells. A **cell** is the smallest unit of a living thing that can carry on life processes such as growing, responding, and reproducing itself. A typical animal cell is shown here. This is a normal cell, not a cancer cell.

An Animal Cell

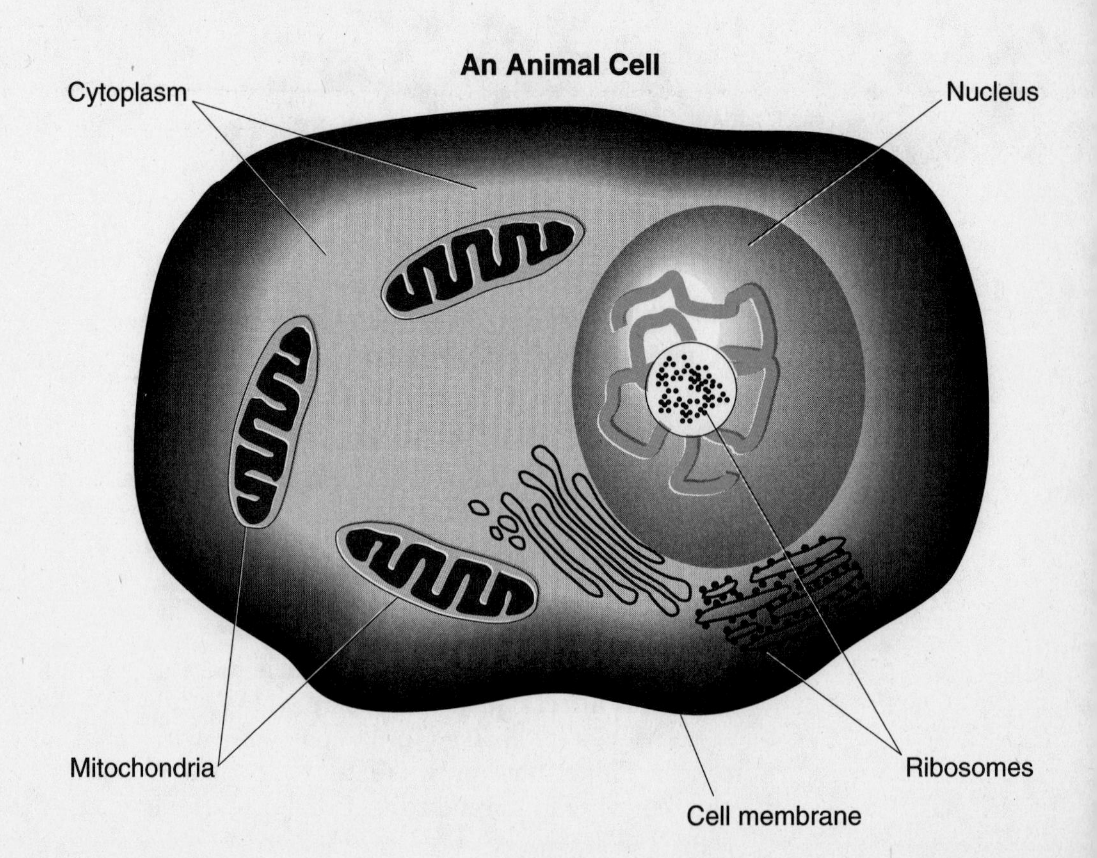

Check your answers on page 213.

A cell has many parts. A jellylike material, called **cytoplasm**, makes up most of the cell. **Ribosomes** make the proteins the cell needs to grow. **Mitochondria** give the cell the energy it needs to grow and reproduce.

The part of the cell that controls all its activities is the **nucleus**. When a normal cell reproduces, it divides into two cells. Each new cell has its own complete copy of the nucleus. The ultraviolet rays in sunlight can cause changes in the nucleus of a skin cell. When this happens, the cells divide abnormally. Cancer is cell division that is out of control.

The **cell membrane** covers the cell. Most cells stop dividing when they touch another cell. Cancer cells keep dividing, even if they crowd the cells near them.

Applying Your Skills and Strategies

Using Headings. Another way to find information is to read the headings in an article. Headings give you an overview of what the material is about. Headings are phrases printed in large, bold, or colored type, so you can see them easily. Write the heading for the part of the article that explains what causes skin cancer.

Preventing Skin Cancer

The color of your skin has a lot to do with how likely you are to get skin cancer. The pigments, or coloring matter, in the skin protect the skin from the sun. In general, people with light skin are the most likely to get skin cancer. People with darker skin, including most Asians and Hispanics, are less likely to get skin cancer. African Americans are least likely to get skin cancer.

Here are some things you can do to protect yourself from skin cancer.

- Spend less time in the sun, especially between 10 A.M. and 3 P.M.
- Wear a sunscreen with an SPF (Sun Protection Factor) of at least 15. Apply it to all of the exposed areas of your body, including the tops of your ears and your lips.
- Use a sunscreen even on cloudy days. Clouds only block some of the ultraviolet light.
- Do not use tanning parlors.
- Check your skin for new growths or sores that do not heal. See a doctor right away. Early treatment for skin cancer is very important.

Check your answer on page 213.

Thinking About the Article

Fill in the blank with the word or words that best complete each statement.

1. ______________________, a type of light in sunshine, can cause skin cancer.
2. ______________________ rarely spreads to other parts of the body.
3. The most dangerous type of skin cancer is called

   ______________________ .
4. The smallest unit of a living thing that can carry on life processes is

   called a ______________________ .
5. The part of the cell that controls all its activities is the

   ______________________ .
6. ______________________ occurs when skin cell division gets out of control.

Write your answers in the space provided.

7. Review the questions you wrote on page 20. Did the article answer your questions? If you said *yes,* write the answers. If your questions were not answered, write three things you learned from this article.

____

____

____

____

____

8. According to the article, what causes skin cancer?

____

9. Reread the paragraph under the heading *Types of Skin Cancer* that discusses basal cell skin cancer. Write one fact that you learned as you read this paragraph.

____

Check your answers on page 213.

10. Reread the paragraph on page 23 about the nucleus of a cell. What happens when ultraviolet rays cause changes in skin cells?

____

11. What are the three types of skin cancer?

____

____

Circle the number of the best answer.

12. Which of the following actions will not protect you from getting skin cancer?
    (1) staying out of the sun at midday
    (2) wearing a sunscreen with an SPF of 15
    (3) using a sunscreen on cloudy days
    (4) going to a tanning parlor weekly
    (5) seeing a doctor about new skin growths

Write your answers in the space provided.

13. Many skin cancers form on the face. Suggest a reason why the face is more likely to be affected than other parts of the body, such as the legs.

____

____

____

14. Read the list of ways to protect yourself. List the two safety measures you are most likely to take.

____

____

15. Describe an experience you or someone you know has had with sunburn or cancer.

____

____

____

Check your answers on page 213.

Blood Vessels

Setting the Stage

Many people have been trying to change their eating habits. They are eating foods with less cholesterol and fat. A diet low in cholesterol and fat helps prevent heart disease.

Past: What you already know

You may already know something about cholesterol, fat, or heart disease. Write two things you already know.

1. ____________________

2. ____________________

Present: What you learn by previewing

Write the headings from the article on pages 27–29 below.

Eating Right for a Healthier Heart

3. ____________________

4. ____________________

5. ____________________

6. ____________________

What do the photos on page 28 show?

7. ____________________

____

Future: Questions to answer

Write two questions you expect this article to answer.

8. ____________________

9. ____________________

 Check your answers on page 213.

Eating Right for a Healthier Heart

As you read each section, circle the words you don't know. Look up the meanings.

Food labels can be confusing. They are full of words like *cholesterol*, *saturated fat*, *polyunsaturated fat*, and *monounsaturated fat*. Since eating too much fat and cholesterol can cause heart disease, it's important to know what's in your food.

Help with Reading the Labels

Many people confuse cholesterol and fat, although these substances are not the same. **Fats** are substances that provide energy and building materials for the body. Fats are found in oils, butter, milk, cheese, eggs, meat, and nuts. When the body takes in more food than it needs, it stores the extra food as fat. **Cholesterol** is a fatlike substance found in all animals, including humans. Some foods, such as egg yolks and shellfish, contain cholesterol. But most of the cholesterol in our bodies is made from the saturated fats we eat.

Saturated fat is a type of fat that is solid at room temperature. Most saturated fats come from animal products. Butter, cheese, meat, egg yolks, and lard have saturated fat. Some vegetable oils, such as palm oil, also have saturated fat. Saturated fat increases the amount of cholesterol in the blood.

Monounsaturated fat is a type of fat found in some vegetable products. This fat is found in olive oil, peanut oil, and peanut butter. Some scientists think that monounsaturated fat lowers the body's cholesterol. Others believe it has no effect on cholesterol.

Polyunsaturated fat is a type of fat found in some vegetable foods and fish. Corn oil, almonds, mayonnaise, soybean oil, and fish are common sources of this fat. Polyunsaturated fat lowers the amount of cholesterol in the blood.

These foods are high in fat and cholesterol.

The Effect of Fat and Cholesterol

Your body needs fat. Fat insulates the body from hot and cold. Fat stores energy. The body uses fats to build cells. Fats are needed to absorb certain vitamins. Women need fat to help regulate menstruation.

Your body also needs cholesterol. Cholesterol is an important part of all animal cells. It also helps protect nerve fibers. The body needs cholesterol to make vitamin D and other substances.

Applying Your Skills and Strategies

Using the Index. Is the word *cell* familiar to you? Cells were mentioned in the article on skin cancer. You can find other discussions of cells in this book by using the index on page 244. An index lists words alphabetically with the page numbers on which the words appear. Find the word *cell* in the index. List three page numbers where cells are discussed.

____

____

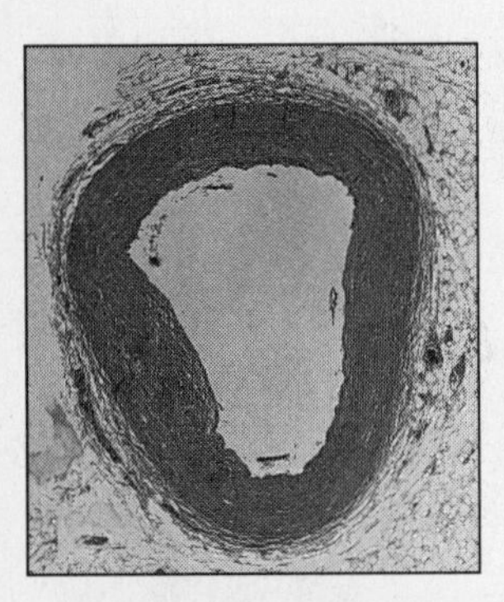

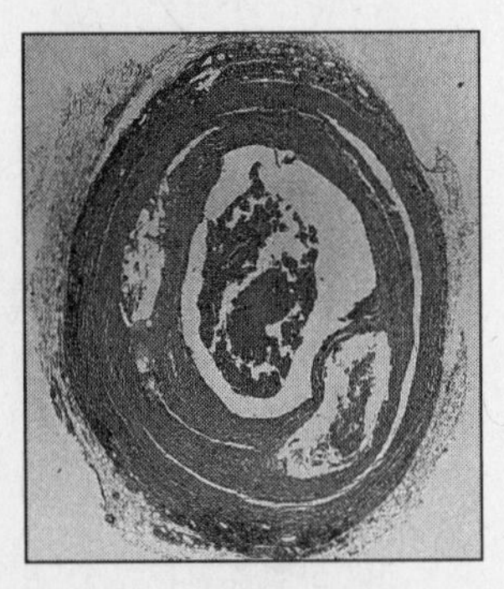

Cross sections of a healthy artery (top) and an artery clogged with plaque (bottom)

If fat and cholesterol have so many beneficial uses, what is the problem? The problem is too much fat and too much cholesterol can be harmful. People should have about one tablespoon of fat in their diet each day. Yet most Americans eat the equivalent of six to eight tablespoons of fat each day! Also, the body can make as much cholesterol as it needs. People don't need to eat cholesterol at all.

What happens when there is too much fat and cholesterol in the diet? Extra fat is stored, and people become overweight. Extra cholesterol circulates in the blood. There it forms deposits called **plaque** on the inside walls of arteries. **Arteries** are large blood vessels that carry blood to all parts of the body. When they are clogged with plaque, arteries cannot carry as much blood. The heart must work harder to pump the same amount of blood. When the flow of blood to the heart muscle is blocked, a heart attack occurs. Part of the heart may die. The person may die.

Studies have shown that people with a lot of cholesterol in their blood are most likely to have heart attacks. Reducing blood cholesterol can lower the risk of having a heart attack. The best way to reduce cholesterol is to eat less food that has cholesterol and saturated fats.

Food Companies Respond

Since so many people are looking for foods low in cholesterol and saturated fat, food companies have responded. Many have switched from saturated to polyunsaturated fats. Many have cut the fat content of their products. Still, you have to read the labels carefully. "No cholesterol" does not mean "no saturated fat."

Check your answer on page 213.

Fast-food companies are changing, too. One major fast-food company stopped frying potatoes in beef fat, which is saturated. Now they use a mixture made mostly of vegetable oil. The food chain also added a salad bar and introduced a low-fat hamburger.

In the new low-fat hamburger, water replaces some of the fat. About 10 percent of the calories comes from saturated fat. In contrast, about 20 percent of the calories from their regular hamburger comes from saturated fat. The two hamburgers are compared here.

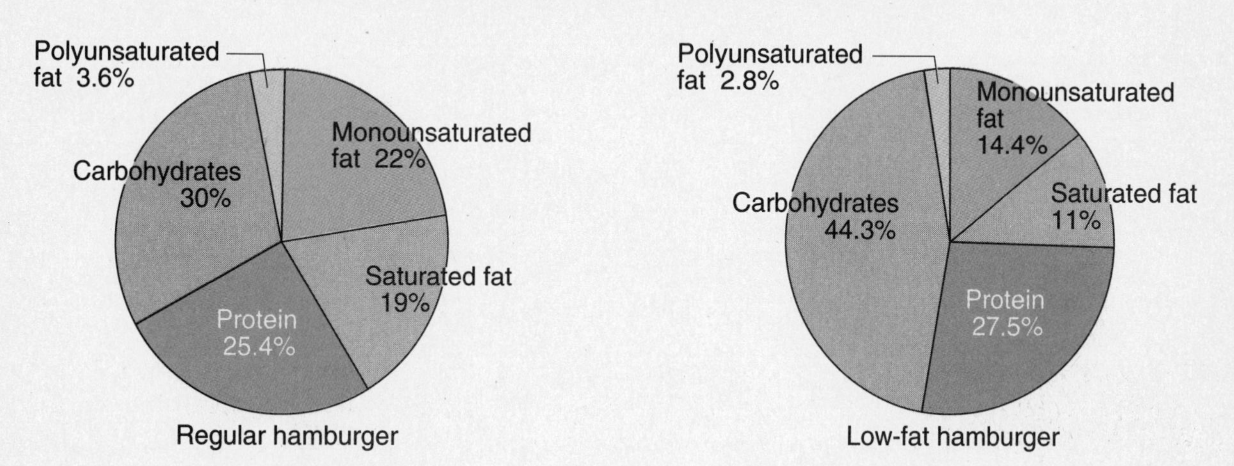

Applying Your Skills and Strategies

Reading a Circle Graph. Think of a circle graph as a pie. The circle is the whole pie. Each wedge is a piece of the pie. In the circle graphs above, the whole pie is the total percent of calories in each hamburger (100%). Each wedge shows the percent of calories for one nutrient. Big wedges show big percents, and small wedges show small percents. Look at the circle graphs to answer these questions.

What percent of the calories in the regular hamburger comes from saturated fat? ______________________________

Which hamburger has more calories from carbohydrates?

____

What You Can Do

There are some simple things you can do to reduce your chances of heart disease. Start with your diet. Eat less meat, butter, ice cream, cheese, and whole milk. Limit eggs to no more than three per week.

Have your cholesterol level checked by your doctor. If it is too high, your doctor will suggest a diet. Get more exercise. Walking, running, swimming, and other aerobic exercises all help lower the body's cholesterol level.

Check your answers on page 214.

Thinking About the Article

Fill in the blank with the word or words that best complete each statement.

1. ______________________ is a type of fat that is solid at room temperature.

2. ______________________ from vegetable foods and fish helps lower the body's cholesterol level.

3. Blood vessels that carry blood to all parts of the body are called

   ______________.

4. ______________ is a deposit of cholesterol on the inside of an artery.

Write your answers in the space provided.

5. Review the questions you wrote on page 26. Did the article answer your questions? If you said *yes*, write the answers. If your questions were not answered, write three things you learned from this article.

____

____

____

____

____

6. Which of the three types of fat raises the body's cholesterol level?

____

7. Where does most of the cholesterol in the body come from?

____

____

8. What happens to extra cholesterol that the body cannot use?

____

____

____

Check your answers on page 214.

Circle the number of the best answer.

9. According to the article, which of the following will help reduce a person's risk of heart disease?

    (1) eating more cheese

    (2) eating fewer eggs

    (3) using butter instead of margarine

    (4) eating fried foods

    (5) drinking whole milk

10. A heart attack happens when

    (1) the arteries send blood to the heart.

    (2) the arteries take blood away from the heart.

    (3) an artery is blocked and not enough blood reaches the heart.

    (4) cholesterol travels around the body in the blood.

    (5) saturated fats cannot be digested.

Write your answers in the space provided.

11. Avocados contain a large amount of saturated fat. Why is this unusual?

____

____

12. Describe an experience you or someone you know has had with heart disease or dieting. Write your description in complete sentences.

____

____

____

____

____

13. If your doctor told you to reduce the amount of saturated fat in your diet, which foods would you be most willing to stop eating? Which foods would be hardest for you to give up?

____

____

Check your answers on page 214.

The Reproductive System

Setting the Stage

Today more babies are born healthy than ever before. One reason is that pregnant women are taking better care of themselves. They know that a woman's health and habits affect her baby's well-being.

Past: What you already know

You may already know something about pregnancy or how unborn babies develop. Write three things you already know.

1. ______________________________

2. ______________________________

3. ______________________________

Present: What you learn by previewing

Write the headings from the article on pages 33–35 below.

Having a Healthy Baby

4. ______________________________

5. ______________________________

6. ______________________________

What does the diagram on page 34 show?

7. ______________________________

Future: Questions to answer

Write three questions you expect this article to answer.

8. ______________________________

9. ______________________________

10. ______________________________

Check your answers on page 214.

Having a Healthy Baby

As you read each section, circle the words you don't know. Look up the meanings.

Scientists once thought that almost all birth defects were **hereditary**. This meant that the problem was passed from parent to child through the sperm or the egg. People did not think that what a mother did could affect the health of her unborn child.

Then in the 1950s and 1960s, many pregnant women in Europe took a drug called *Thalidomide*. Thousands of these women had babies born with misshapen arms and legs. Over a thirty-year period in this country, millions of women took a drug called *DES*. This drug prevented miscarriages. But by 1970, daughters of many of these women had cancer. Studying the effects of these two drugs helped scientists focus their research. Since then, scientists have learned that a baby is affected by what its mother eats, drinks, and smokes.

A mother and her healthy baby

The Development of an Unborn Baby

In nine months, a new human being develops from one cell into an organism with billions of cells. A developing baby goes through three stages before it is born. First it is a zygote, then an embryo, and finally a fetus.

Zygote. A sperm from the father joins with the egg produced by the mother. This fertilized egg is called a **zygote**. The zygote divides and forms a hollow ball of cells. By the tenth day, the zygote attaches to the **uterus**, or womb. During this early stage, there is no exchange of substances between the mother and the zygote.

Embryo. From the third to the eighth week, the developing baby is called an **embryo.** The embryo is attached to the uterus by the placenta. The **placenta** is a structure that allows substances to pass back and forth between the embryo and the mother.

The placenta is like a filter. The mother's blood passes along one side of the placenta. The embryo's blood passes along the other side. The two bloodstreams are separated by a thin layer of cells. Nutrients and oxygen pass from the mother's blood to the embryo's. Waste and carbon dioxide pass from the embryo's blood to the mother's. Other substances in the mother's blood, such as drugs or chemicals, can pass through, too.

During this stage, the main body organs and systems form. Birth defects are most likely to occur at this time. A harmful substance in the mother's blood can cause serious damage to the embryo.

Fetus. From the third to the ninth month, the developing baby is called a **fetus**. During these months, the fetus grows. Harmful substances are less likely to cause birth defects during this stage because the main systems and body have been formed. If damage occurred in the embryo stage, the damaged organ or system in the fetus may not work properly.

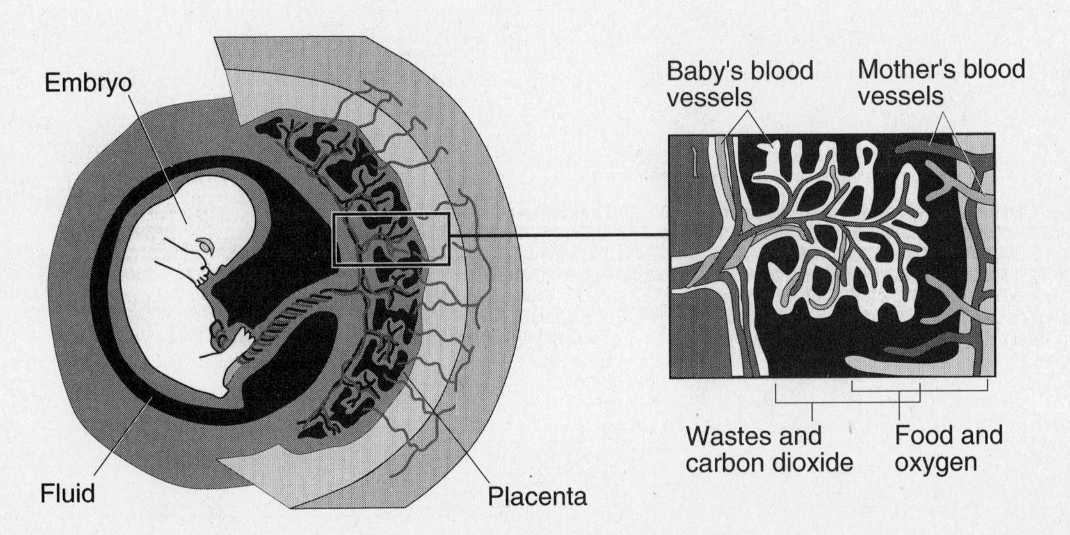

The Exchange of Substances Between Mother and Embryo

Applying Your Skills and Strategies

Understanding a Diagram. Sometimes words do not communicate as well as a diagram does. A diagram helps you see how something looks or works. The main idea of a diagram is often stated in its title. What is the main idea of the diagram above?

____

____

The labels or words that identify parts of the diagram are important. They tell you what to look at. Circle all of the labels in the diagram above.

Check your answers on page 214.

The Effects of Drugs

What happens when a pregnant woman smokes a cigarette, drinks a beer, or takes an aspirin? It's hard to prove what is harmful and what is not. But over the years, scientists have started to link some substances with certain problems.

Cocaine. Studies show that in some cities as many as one in ten babies have been exposed to cocaine before birth. The mother's use of cocaine can cause a miscarriage, early labor, or stillbirth. Cocaine babies are often underweight. They may have damage to the brain, lungs, urinary system, or sex organs.

Cigarettes. The nicotine in cigarettes makes the blood vessels in the placenta shrink. Less oxygen and fewer nutrients reach the developing baby. Smoking has been linked to miscarriages and stillbirths. Babies born to smokers are often underweight at birth.

Applying Your Skills and Strategies

Finding Details. The main idea of a paragraph is what the paragraph is generally about. The details are facts that describe or explain the main idea. For example, the main idea of the previous paragraph is that smoking during pregnancy can affect the baby. A detail that supports this main idea is that the babies of smokers often are underweight at birth. Find two other details in this paragraph.

Alcohol. **Fetal alcohol syndrome (FAS)** can occur when the mother drinks alcohol while pregnant. Children with FAS may have abnormal heads, faces, arms, or legs. These children often have low birth weights. Some children with FAS are mentally retarded.

Common Medications. Most common medications seem to pose no risk to the developing baby. This does not mean there are no risks. It just means no risks have been proved. However, some risks are suspected. Taking a lot of aspirin may cause bleeding or longer labor. Some aspirin substitutes taken late in pregnancy may cause problems with the baby's breathing.

Preventing Birth Defects

There is no way to guarantee a healthy baby. But women can do some things to help lower the risks to their babies. A pregnant woman should eat a well-balanced diet. She should get regular care from a doctor. A woman who is pregnant should not smoke or drink alcohol. She should not take cocaine or other illegal drugs. Also, she should check with her doctor before taking any medicines.

Check your answers on page 214.

Thinking About the Article

Fill in the blank with the word or words that best complete each statement.

1. The fertilized egg is called a ____________.
2. From the third to the eighth week, the developing organism is called an ____________.
3. The ______________ is a structure that allows substances to pass back and forth between the unborn baby and the mother.

Write your answers in the space provided.

4. Review the questions you wrote on page 32. Did the article answer your questions? If you said *yes*, write the answers. If your questions were not answered, write three things you learned from this article.

____

____

____

____

____

5. Which two drugs helped focus research on what happens when a woman takes drugs during pregnancy?

____

6. During which stage of development is the most serious damage to the unborn baby likely to occur?

____

Circle the number of the best answer.

7. Brain damage in babies has been linked with the mother's use of
   (1) aspirin.
   (2) aspirin substitutes.
   (3) cigarettes.
   (4) cocaine.
   (5) all of the above.

 Check your answers on page 214.

8. A pregnant woman who wants to have a healthy baby should
   (1) eat a well-balanced diet.
   (2) stop smoking if she smokes.
   (3) stop drinking alcohol if she drinks.
   (4) use drugs only with the advice of a doctor.
   (5) do all of the above.

9. Harmful substances are least likely to cause birth defects in the last few months of pregnancy. The reason for this is that, by the last few months of pregnancy,
   (1) all of the body organs have formed.
   (2) the placenta blocks harmful substances.
   (3) the fetus has stopped growing.
   (4) the fetus can control what it takes in.
   (5) the fetus is not connected to the placenta.

Write your answers in the space provided.

10. Describe an experience you or someone you know has had with pregnancy or babies. Write your description in complete sentences.

____

____

____

____

11. Many states have laws that require alcoholic beverages to have a warning label stating the dangers of drinking while pregnant. Do you think this law works well? Give a reason for your answer.

____

____

____

12. If you had a friend who was pregnant and you saw her lighting a cigarette, what would you do? Give a reason for your answer.

____

____

Check your answers on page 214.

Section 5

Bones and Muscles

Setting the Stage

One of the simplest activities—walking—is good for your health. It's a low-impact aerobic exercise that strengthens the heart, muscles, and bones. Best of all, walking doesn't require expensive gear, special clothing, or lessons.

Past: What you already know

You may already know something about walking for pleasure or fitness. Write three things you already know.

1. ______________________________

2. ______________________________

3. ______________________________

Present: What you learn by previewing

Write the headings from the article on pages 39–41 below.

Walking for Fitness

4. ______________________________

5. ______________________________

6. ______________________________

What do the diagrams on page 40 show?

7. ______________________________

Future: Questions to answer

Write three questions you expect this article to answer.

8. ______________________________

9. ______________________________

10. ______________________________

Check your answers on page 215.

Walking for Fitness

As you read each section, circle the words you don't know. Look up the meanings.

Strolling through the mall or around town is a pleasant way to spend time. But did you know that by walking faster you can improve your fitness and health? Walking at a fast pace gives the heart, lungs, muscles, and bones a good workout.

The Benefits of Walking

Walking is easy and it's convenient. You don't need expensive equipment, special clothing, or lessons to walk. All you need is a good pair of shoes and a little time. You can walk anywhere—around the block, on a track, or even indoors. Some malls open early so walkers can exercise before the shoppers arrive.

Walking is easy on the body. It doesn't jar the joints as jogging or tennis do. That makes walking safe for many people. In fact, doctors often recommend walking to patients recovering from heart attacks, operations, or injuries.

Walking is a safe and inexpensive way to exercise.

Walking is an aerobic exercise like jogging, swimming, cycling, or aerobic dancing. Walking strengthens the heart by making it beat faster and harder. Walking also increases the amount of oxygen taken in by the lungs. Studies have shown that brisk walking may reduce the risk of heart disease and high blood pressure.

Walking uses up stored fat. Walking burns about the same number of calories per mile as jogging. It just takes longer to walk a mile than to jog. People who walk regularly and do not increase the amount they eat can lose weight slowly.

Improving Bones and Muscles

Walking strengthens major muscle groups in the legs, abdomen, and lower back. Walkers who pump their arms can make the muscles of the upper body stronger, too.

Muscles work in pairs to move bones. As you push off the ground, you straighten your ankle. To do this, the calf muscle of the leg contracts, or shortens. At the same time, the shin muscle relaxes, or lengthens. Then as you swing your leg forward, your foot lifts up. To bend the ankle this way, the shin muscle contracts. At the same time, the calf muscle relaxes. This repeated movement improves the tone of the leg muscles.

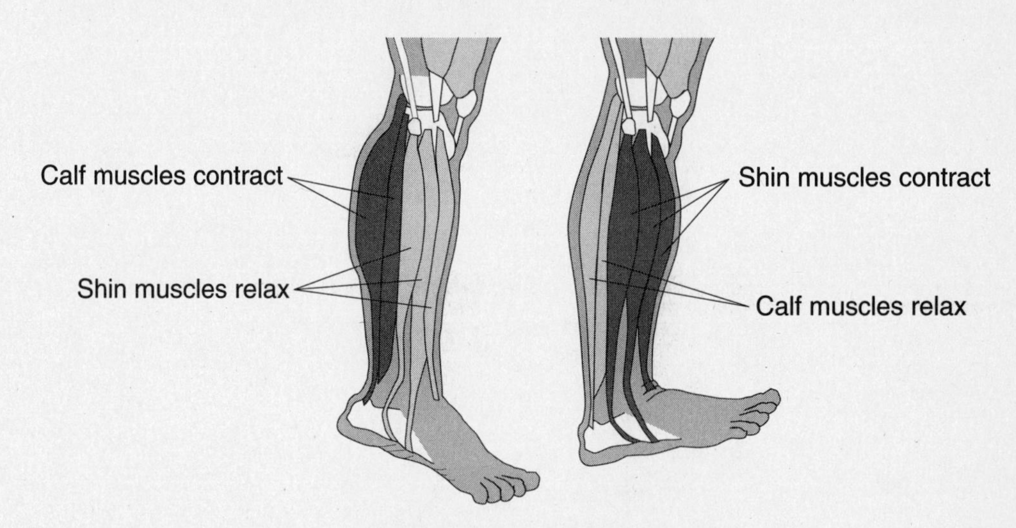

Muscles Work in Pairs

Exercise that increases muscle strength also benefits the bones. Exercise causes muscles to pull on the bones. As a result, the bones become stronger and more solid. This can help prevent **osteoporosis**, a condition of brittle bones. This problem affects many older people, especially women.

Applying Your Skills and Strategies

Finding the Implied Main Idea. Sometimes the main idea of a paragraph is not stated. Instead, it is hinted at, or *implied*. You have to read between the lines to get the general idea. Put all of the details together to get the main idea. Reread the paragraph on this page that begins, "Exercise that increases. . . ." It discusses the effect of exercise on bones. The implied main idea is that ***walking can benefit bones.*** Notice that the word *walking* is never mentioned.

Reread the first paragraph on this page. Write the implied main idea of that paragraph.

____

____

 Check your answer on page 215.

Easy on the Joints

Walking is also good exercise because it doesn't jar the joints of the body. A **joint** is a place where two or more bones come together. There are several types of joints.

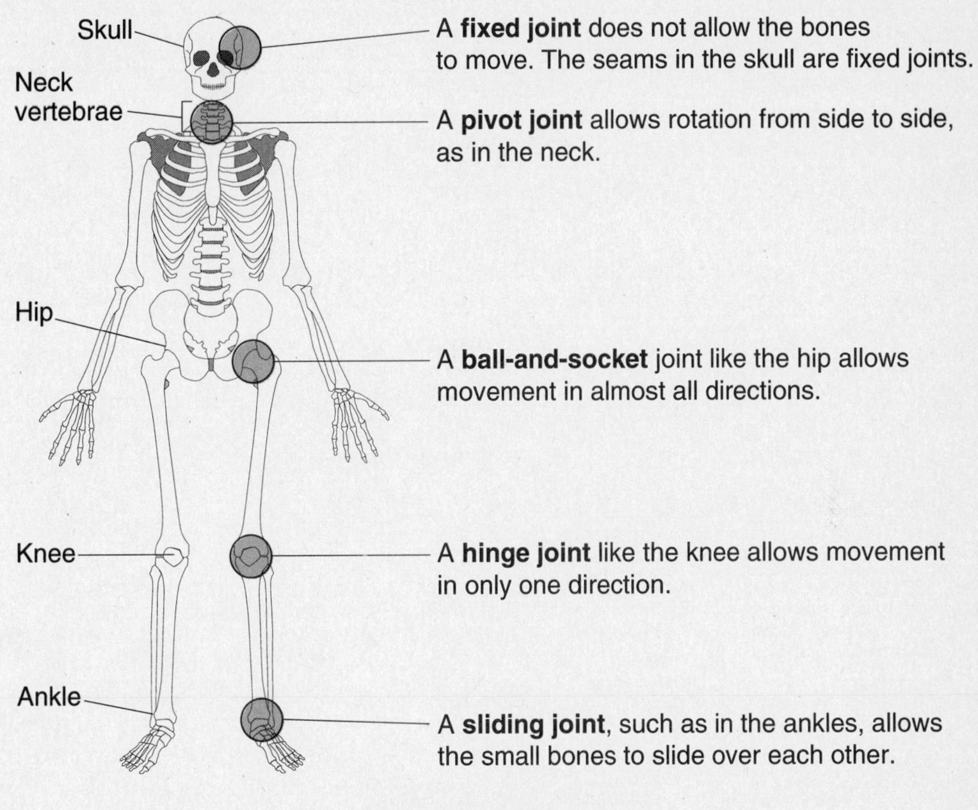

The Joints of the Body

Jogging can put a lot of stress on joints. Sometimes runners get a sprain. A **sprain** is a joint injury in which the ligaments are stretched or torn. **Ligaments** are strong bands that connect bones at joints. In contrast, walkers rarely get sprains.

Applying Your Skills and Strategies

Comparing and Contrasting. Sometimes writers explain an idea by telling how it is different from something else. This is called *contrasting*. In the previous paragraph, the writer contrasted walking and running. Running causes more joint injuries than walking. At times writers tell how things are similar. This is called *comparing*. Reread the last paragraph on page 39. Compare the ways that running and walking are similar.

____

____

____

Check your answer on page 215.

Thinking About the Article

Look at the diagram on page 41. Match the type of joint with its description. Write the letter of the joint in the blank to the left.

	Description	Joint
______	1. allows movement in only one direction (knee)	a. ball-and-socket joint
______	2. allows movement in almost all directions (hip)	b. fixed joint
______	3. allows sliding movements (ankle)	c. hinge joint
______	4. allows rotation from side to side (neck)	d. pivot joint
______	5. allows no movement (skull)	e. sliding joint

Write your answers in the space provided.

6. Review the questions you wrote on page 38. Did the article answer your questions? If you said *yes*, write the answers. If your questions were not answered, write three things you learned from this article.

____

____

____

____

____

7. How does walking improve the heart?

____

____

8. Give two reasons why walking is an exercise that almost anyone can do.

____

____

 Check your answers on page 215.

Circle the number of the best answer.

9. Walking for fitness can provide which of the following benefits?
   (1) improved muscle tone
   (2) reduced risk of heart disease
   (3) stronger bones
   (4) burning of fat
   (5) all of the above

10. What happens when you bend your ankle?
    (1) One leg muscle contracts and the other relaxes.
    (2) Both leg muscles contract.
    (3) Both leg muscles relax.
    (4) Ligaments pull the bone up toward the body.
    (5) Ligaments push the bone away from the body.

11. Walking can strengthen the muscles of the upper body. To get this benefit, what must you do as you walk?
    (1) Walk faster.
    (2) Pump your arms.
    (3) Hold in your stomach.
    (4) Hold your head up.
    (5) Focus your eyes on the ground in front of you.

Write your answers in the space provided.

12. Compare and contrast walking and running.

____

____

____

13. Describe an experience you or someone you know has had with walking. Write your description in complete sentences.

____

____

____

Check your answers on page 215.

Bacteria and Viruses

Setting the Stage

How many times in the last year have you had a cold? Most people have at least two colds each year. Even worse than a cold is the flu. Although most people recover, sometimes colds and flu lead to pneumonia.

Past: What you already know

You may already know something about colds, flu, or pneumonia. Write two things you already know.

1. ______________________________

2. ______________________________

Present: What you learn by previewing

Write the headings from the article on pages 45–47 below.

Colds, Flu, and Pneumonia

3. ______________________________

4. ______________________________

5. ______________________________

6. ______________________________

What does the photo on page 45 show?

7. ______________________________

Future: Questions to answer

Write three questions you expect this article to answer.

8. ______________________________

9. ______________________________

10. ______________________________

 Check your answers on page 215.

Colds, Flu, and Pneumonia

As you read each section, circle the words you don't know. Look up the meanings.

First you feel a small ache or a tickle in your throat. Soon your nose is running, your head is congested, and your eyes are watering. Yes, it's another cold.

Catching a Cold

In spite of what your mother may have told you, getting wet and chilled won't give you a cold. Instead, you catch a cold from a sick person near you.

Scientists disagree about how colds are passed from one person to another. Jack Gwaltney, a scientist at the University of Virginia, thinks that colds are passed by touch. Gwaltney says that people with colds have many cold-causing viruses on their hands. When these people touch something, such as a telephone, they leave viruses on the surface. You come along and touch the telephone. Then you touch your nose or eyes, and the virus settles in.

Another scientist, Elliot Dick from the University of Wisconsin, disagrees with Gwaltney. As an experiment, Dick gathered sixty card players for a twelve-hour poker game. Twenty players had colds. Of the forty healthy players, half wore braces or collars that kept them from touching their faces. The other half were free to touch their faces. Players in both healthy groups caught colds. Since players who could not touch their faces also got sick, Dick believes that cold viruses are spread through the air. Someone who is sick sneezes or coughs, and you breathe in the virus. Soon you are sick, too.

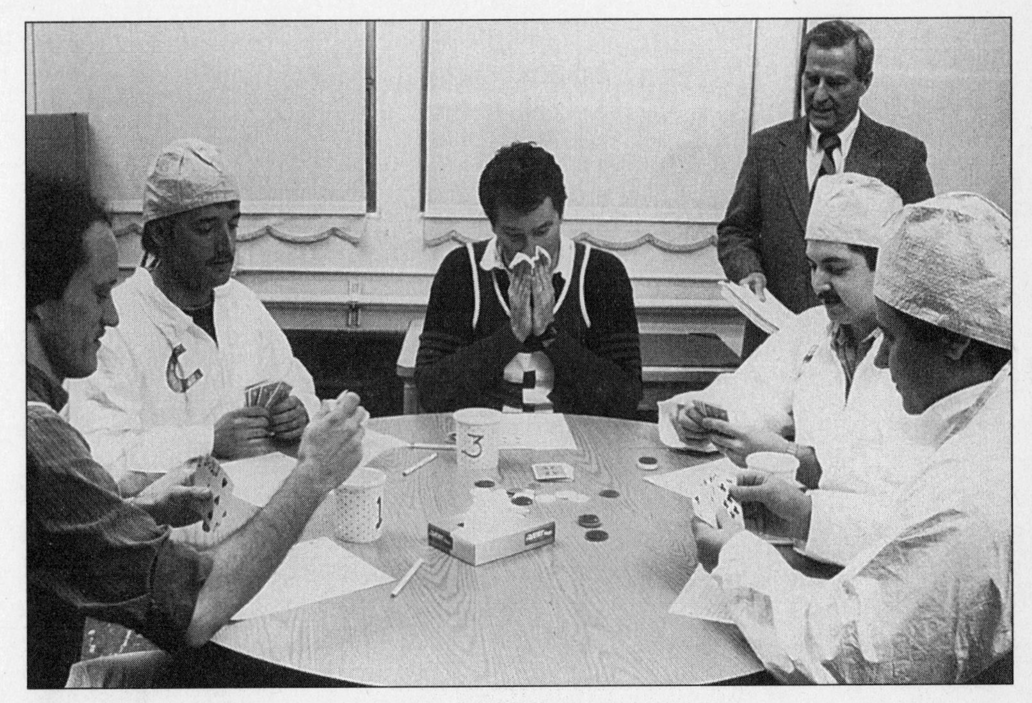

Elliot Dick's cold virus experiment

Treating a Cold

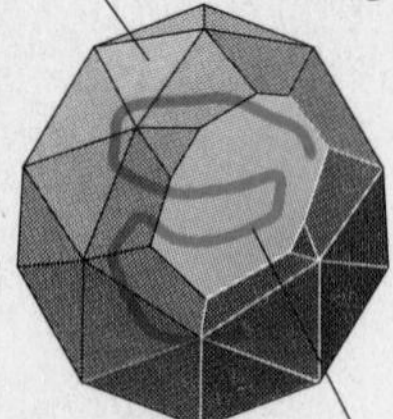

A Rhinovirus

Stroll through any drugstore and you'll see hundreds of sprays, pills, capsules, and syrups to make the cold sufferer feel better. Many of these products do help. But none can cure a cold. For years, scientists have been trying to find a cure for the common cold. The problem is that colds are caused by about 200 different viruses. About half of all colds are caused by a group of viruses called **rhinoviruses**.

Just what is a virus? A **virus** is a tiny particle made up of genetic material with a protein covering. The genetic material holds directions for making more viruses. The protein covering protects the virus.

Applying Your Skills and Strategies

Understanding a Diagram. A diagram can help you see what the author is describing. The title tells you what the diagram is about. Circle the title of the diagram on this page. The words with lines pointing to parts of the diagram are called *labels*. Labels direct your attention to important parts of the diagram. Underline the labels in the diagram.

Tell in your own words what the diagram shows.

____

____

Viruses are neither living nor nonliving things. Viruses are not cells, and they are not made of cells. They don't grow. To reproduce, they must be inside a living cell. For example, a rhinovirus latches onto a cell in the nose. The virus injects its genetic material into the cell and uses the cell's materials to make more viruses. Then the cell bursts open and dies. The new viruses are released to infect other cells.

Infections caused by viruses are hard to cure. The best way to fight viruses is to stop them before they invade. Researchers are looking for ways to block the rhinoviruses from attaching to cells in the nose.

The Flu

Another illness caused by a virus is **influenza**, usually called the **flu**. The difference between a cold and the flu is sometimes hard to feel. A cold usually starts slowly, with a scratchy throat. Other symptoms follow in a day or two. Usually there is little or no fever. In contrast, the flu starts quickly. Within twelve hours there may be a fever over 101 degrees. Besides giving you a sore throat, stuffy nose, and a cough, the flu makes you ache and feel very tired. Recovering from the flu can take one to three weeks. Although there are some drugs to help people feel better, there is no cure for the flu.

Check your answer on page 216.

The best defense against the flu is to get a flu vaccination each year. A **vaccination** is an injected dose of a dead or weakened disease-causing agent. This causes the body to form antibodies. An **antibody** is a substance the body makes to fight a disease. Once antibodies form, you are protected against that specific agent. Since the flu is caused by only one or two viruses, vaccinations have been successful.

Pneumonia

People weakened by a bad cold or the flu are more likely to develop pneumonia. Symptoms include a high fever, chest pain, breathing problems, and a bad cough. **Pneumonia** is an infection of the lungs caused by a virus or bacteria. Unlike viruses, **bacteria** are living things. They are one-celled organisms and are many times larger than viruses. A bacterial cell is shown here.

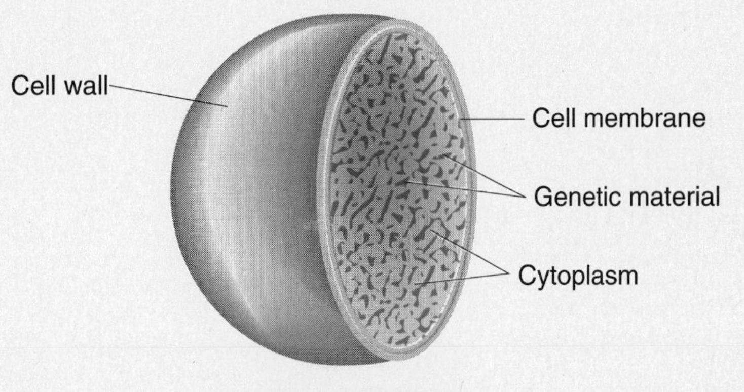

A Bacterial Cell

About 90 percent of flu-related pneumonia is caused by bacteria. As the infection takes hold, the lungs produce fluid and mucus to fight the bacteria. The fluid blocks the air passages in the lungs. The person has trouble breathing and may have to go to a hospital. He or she is given **antibiotics**, drugs that fight bacteria.

Applying Your Skills and Strategies

Making Inferences. As you read, look for main ideas and details. Sometimes you can figure out things that the author hints at but doesn't actually tell you. A fact or idea you figure out that is not stated in the text is an *inference*. Reread the paragraphs about the flu and pneumonia. The author says that people who have the flu sometimes develop pneumonia. One inference you can make is that a good way to avoid getting pneumonia is to avoid getting the flu. Write another inference about the flu or pneumonia.

____

____

Check your answer on page 216.

Thinking About the Article

Fill in the blank with the word or words that best complete each statement.

1. A tiny particle, neither living nor nonliving, that is made up of genetic material and a protein covering is called a ____________.
2. ________________________, which is caused by a virus, has symptoms of high fever, aches, and tiredness.
3. ___________________ is an infection of the lungs caused by bacteria or viruses.
4. ___________________ are a type of one-celled living thing.
5. Drugs called _________________ are used to fight bacterial infections.
6. An _______________________ is a substance the body makes to fight a disease.

Write your answers in the space provided.

7. Review the questions you wrote on page 44. Did the article answer your questions? If you said *yes*, write the answers. If your questions were not answered, write three things you learned from this article.

____

____

____

____

____

8. According to the diagram on page 47, what are the parts of a bacterial cell?

____

____

____

____

Check your answers on page 216.

Circle the number of the best answer.

9. Bacteria can cause
   (1) colds.
   (2) flu.
   (3) pneumonia.
   (4) flu and pneumonia.
   (5) colds, flu, and pneumonia.

10. Rhinoviruses can cause
   (1) colds.
   (2) flu.
   (3) pneumonia.
   (4) flu and pneumonia.
   (5) colds, flu, and pneumonia.

11. Antibiotics are not used to treat colds and the flu because
   (1) they do not fight viruses.
   (2) they do not fight bacteria.
   (3) people with colds or the flu do not go to the hospital.
   (4) they cost too much to use them on colds and the flu.
   (5) they are used to prevent colds and the flu.

Write your answers in the space provided.

12. Turn back to page 41 and review the skill of comparing and contrasting. Then compare and contrast the symptoms of a cold and the flu.

____

____

____

13. Describe an experience you or someone you know has had with a cold or the flu.

____

____

____

Check your answers on page 216.

Section 7

Genetics

Setting the Stage

Do you have the same color eyes or hair as your mother? Parents pass on features like these to their children. They can also pass on certain diseases. Scientists have developed tests to see if people are likely to get a disease passed on by their parents.

Past: What you already know

You may already know something about characteristics that are passed from parents to their children. Write three things you already know.

1. ______________________________

2. ______________________________

3. ______________________________

Present: What you learn by previewing

Write the headings from the article on pages 51–53 below.

Genetic Screening

4. ______________________________

5. ______________________________

6. ______________________________

What does the diagram on page 52 show?

7. ______________________________

Future: Questions to answer

Write three questions you expect this article to answer.

8. ______________________________

9. ______________________________

10. ______________________________

 Check your answers on page 216.

Genetic Screening

As you read each section, circle the words you don't know. Look up the meanings.

When Kristi Betts was fifteen, her mother came down with **Huntington's disease**. This disorder, which starts in middle age, slowly attacks the brain. People with Huntington's lose control over their physical and mental functions. This takes place over a period of about twenty years. There is no treatment or cure.

At first, Kristi Betts was concerned only for her mother. Then she realized that she and her sisters and brothers were at risk for Huntington's disease, too. You cannot catch Huntington's disease like a cold or the flu. You **inherit**, or receive, the disease from a parent. Huntington's disease is passed on in the same way as brown eyes or curly hair. Each child of a parent with Huntington's disease has a fifty-fifty chance of inheriting it.

How Traits Are Passed from Parent to Child

All organisms inherit characteristics, or **traits**, from their parents. Some traits, such as hair color, are easily seen. Others, such as blood type, cannot be seen. **Heredity** is the passing of traits from parents to their young, or offspring. Traits are passed on when organisms reproduce. The study of how traits are inherited is called **genetics**.

Each parent gives its offspring a form of the trait. In some cases, one form of the trait shows and the other does not. The form of the trait that shows is called the **dominant trait**. The form of the trait that doesn't show is called the **recessive trait.** In humans, for example, dark hair and dark eyes are dominant traits. Light hair and light eyes are recessive traits.

A four generation birthday party

Scientists show how traits combine by using a diagram called a Punnett square. A **Punnett square** shows all the possible combinations of traits among the offspring of two parents. Look at the Punnett square on the left. It shows the combinations that may result when one parent has the trait for Huntington's disease and the other parent does not. The trait for Huntington's disease is a dominant trait. It is labeled with a capital H. The healthy form of the trait is recessive. It is labeled with a lowercase h.

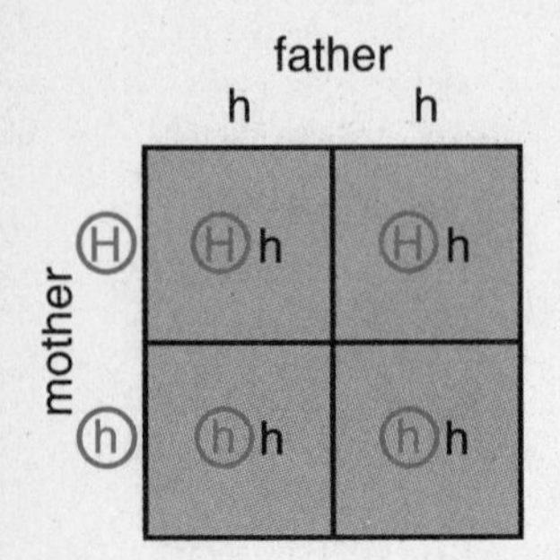

A Punnett Square

In this Punnett square, the mother with the trait for Huntington's disease is shown on the left. She has a dominant and a recessive form of the trait. The healthy father is shown on the top. He has two recessive forms of the trait. When these two people have a child, each gives one form of the trait to the child. Their traits can combine as shown in the four boxes. The child may inherit a dominant and recessive form (Hh). He or she will develop Huntington's disease as an adult. A child who inherits two recessive forms (hh) will not develop the disease.

Applying Your Skills and Strategies

Finding Details. Details are very important in science. When you come to a passage with a lot of details, slow down. The previous paragraph explains how to read the Punnett square. Reread the paragraph one sentence at a time. If you read slowly and carefully, you will understand the details. If you understand the details, then you'll understand the main idea, too.

On the Punnett square shown, what does the h stand for?

____

____

Will a child who inherits hh develop Huntington's disease?

____

____

Testing for Inherited Disorders

You can see from the Punnett square above that Kristi Betts had a two out of four chance of having the dominant form of the Huntington's trait. At one time, people had to live with this uncertainty. Since Huntington's disease doesn't appear until middle age, young people with an ill parent didn't know their own fate. But now people at risk can be tested. **Genetic screening** is the name for tests that can tell people if they have inherited certain disorders.

It is hard to decide whether to take the test for Huntington's disease. Is it better to know or not know that you are going to get the disease? Kristi Betts decided it was better to know. She was lucky. She found out that she had not inherited Huntington's disease.

 Check your answers on page 216.

Many genetic diseases are recessive. When two recessive forms combine, the child inherits the disorder even though both parents may be healthy. There are tests to see if adults carry the recessive traits for certain disorders, such as sickle-cell anemia.

Genetic Screening of the Unborn

Genetic screening of fetuses is becoming more common. Pregnant women can be tested to see whether the fetus has certain disorders. **Amniocentesis** is a test that can be performed after the sixteenth week of pregnancy. This test can detect several disorders, including Down syndrome. **Down syndrome** is a disorder caused by the presence of an extra chromosome. Children born with Down syndrome are mildly to severely mentally retarded. They may also have other health problems.

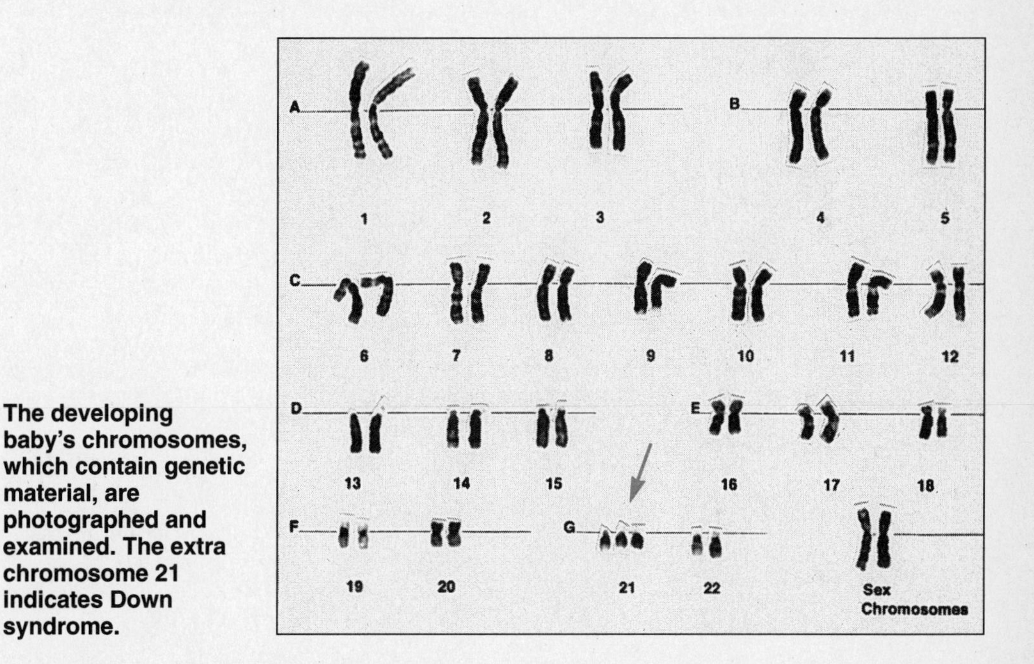

The developing baby's chromosomes, which contain genetic material, are photographed and examined. The extra chromosome 21 indicates Down syndrome.

When test results show a disorder, parents must decide what to do. Genetic counselors help parents explore their choices. Parents may decide not to have children. They may also accept the outcome and prepare for the birth of a disabled child. Some people believe that genetic screening is wrong. Others feel that knowledge, even if it is bad news, is worth having.

Applying Your Skills and Strategies

Distinguishing Fact from Opinion. Facts are things that can be proved true. Opinions, on the other hand, are beliefs. They may or may not be true. When reading science, it's important to distinguish fact from opinion. One way to do this is to look for words like *believe*, *feel*, *think*, and *opinion*. Reread the last paragraph of the article. Circle any words that signal opinions. Then write your own opinion about genetic screening in the space below.

Check your answer on page 216.

Thinking About the Article

Fill in the blank with the word or words that best complete each statement.

1. ______________________________ is an inherited disorder that attacks the brain.
2. __________, or characteristics, are inherited from parents.
3. The passing on of traits from parents to their children is called

   ____________.
4. ____________ is the study of how traits are inherited.
5. In humans, dark hair is a trait that shows up or is ____________.
6. The form of a trait that doesn't show up is called

   ______________.
7. A diagram called a ____________________ is used to show all possible combinations of a trait among the offspring of two parents.

Write your answers in the space provided.

8. Review the questions you wrote on page 50. Did the article answer your questions? If you said *yes*, write the answers. If your questions were not answered, write three things you learned from the article.

____

____

____

____

____

9. In a Punnett square, what does a capital letter represent?

____

____

 Check your answers on page 216.

Circle the number of the best answer.

10. Which of the following is a fetal genetic screening test?
    (1) Huntington's disease
    (2) amniocentesis
    (3) Punnett square
    (4) heredity
    (5) Down syndrome

11. According to the Punnett square on page 52, what are the chances that a child of a parent with Huntington's disease will inherit the disease?
    (1) zero out of four
    (2) one out of four
    (3) two out of four
    (4) three out of four
    (5) four out of four

Write your answers in the space provided.

12. Why might a person whose parent has Huntington's disease decide not to have the genetic test?

____

____

____

13. Think about one of your parents. Write three traits that your parent has passed on to you.

____

____

14. If you or your spouse were going to have a baby, would you want genetic screening during pregnancy? Why or why not?

____

____

____

Check your answers on pages 216-217.

Plants

Setting the Stage

Many medicines have been made from the leaves, roots, stems, flowers, and bark of plants. Many plants that grow in the tropical rain forests are used for medicines. Scientists are working to find these useful plants before the rain forests disappear.

Past: What you already know

You may already know something about plants used for medicine or about the rain forests. Write three things you already know.

1. ______________________________

2. ______________________________

3. ______________________________

Present: What you learn by previewing

Write the headings from the article on pages 57–59 below.

Medicines from Plants

4. ______________________________

5. ______________________________

6. ______________________________

What does the table on page 57 show?

7. ______________________________

____

Future: Questions to answer

Write two questions you expect this article to answer.

8. ______________________________

9. ______________________________

 Check your answers on page 217.

Medicines from Plants

As you read each section, circle the words you don't know. Look up the meanings.

In South America, the juice from a fungus is used to treat earaches. In Indonesia, pressed plants are used to cure tetanus. In India and Nepal, coleus is used to get rid of body worms. These are just a few of the plants that are used as medicines. People have been using plants to treat and cure diseases for thousands of years. Plant medicines work in two ways. Some affect the body's chemistry. Others affect the bacteria and viruses that cause diseases.

Using Plant Medicines

Even in more modern cultures, plants are the source of many drugs. There are more than 265,000 different kinds of plants, or plant species, known in the world. Less than 1 percent of them have been tested as medicines. Yet about 25 percent of all medicines have come from that tiny number of plants. Many medicines are made from natural plant products. Others are synthetic, or made by people. The table shows some medicines and their plant sources.

Medicines and Their Plant Sources

Medicine	Plant Source	Use
quinine	cinchona bark	treating malaria
curare	several tropical plants	muscle relaxant during surgery
digitalis	purple foxglove	treating heart disorders
vinca alkaloids	rosy periwinkle	treating Hodgkin's disease and leukemia
expectorant	horehound	in cough syrup to discharge mucus
menthol	peppermint leaves	in pain relievers and decongestants
valepotriates	root of valerian	sedative
reserpine	roots of several shrubs	sedative; treating high blood pressure
salicylate	wintergreen	in liniments to sooth muscle aches
taxol	bark of Pacific yew	treating ovarian and other cancers

The table shows that many common medicines come from plants. For example, Valium is a well-known sedative. It is a synthetic form of substances found in the roots of the valerian plant. Horehound is used to make cough syrups and throat lozenges.

Some plants are used in the treatment of serious diseases, such as cancer. Many forms of chemotherapy use natural products from plants. A recent discovery is taxol. It is used to treat ovarian cancer. So far, taxol cannot be made in a laboratory. The Pacific yew must be harvested to produce natural taxol.

Parts of a Plant

Medicines have been made from all parts of plants. Often only one part of a plant can be used to make a medicine. The diagram shows the parts of a typical seed plant.

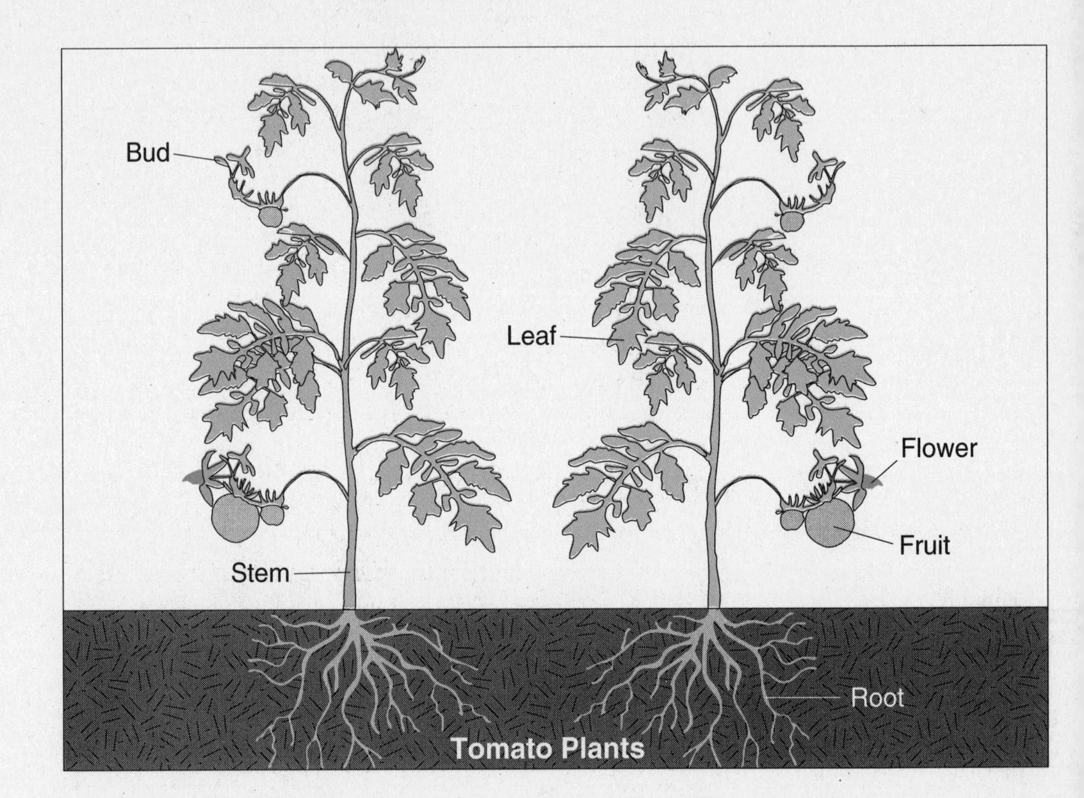

The roots of a plant hold the plant in the ground. They also absorb minerals and water from the soil. The stem supports the plant. It also transports water and minerals from the roots to the top of the plant. Bark is the outer part of the stem of a woody plant. Leaves use sunlight to change water and carbon dioxide into food for the plant. Buds are areas of growth. New branches or flowers grow from buds. Flowers are the reproductive organs of plants. Flowers develop into fruits. The fruits contain the seeds, which will grow into new plants.

Applying Your Skills and Strategies

Getting Meaning from Context. You may not understand every word you read in science. But sometimes you can figure out the meaning of words you don't know. Reread the description of the stem above. Can you guess what the word *transport* means? Look at the context—the rest of the words in the sentence. There are clues. Water and minerals are moving up from the roots to the rest of the plant. The stem is carrying water and minerals. The word *transport* means "to carry or move."

Reread pages 57 and 58 of the article. Circle words you do not know. Try to guess their meanings. Write the words and your guesses on a sheet of paper. Then look up the words in a dictionary.

Check your answers on page 217.

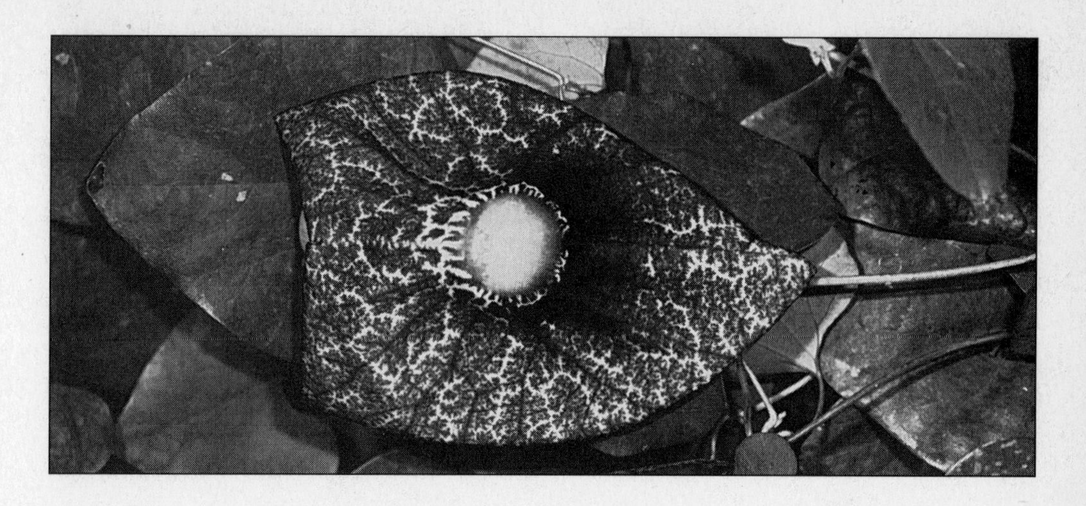

In the rain forest, this vine is brewed into a tea, which is used to treat stomachaches.

Searching for Medicinal Plants

Scientists must test a plant if they think it may be a source for a medicine. An extract, or strong solution, is made from the plant. The extract is placed in test tubes that contain cancer cells, bacteria, or viruses. Extracts that seem to work as a medicine are tested on mice. Then the extract is broken down into its chemical parts. The parts are tested until the active chemical is found. Then scientists must figure out how much of the extract is safe to use. Many plant products are poisonous in large doses.

Applying Your Skills and Strategies

Summarizing Information. When you summarize something you condense, or shorten, a larger amount of information into a few sentences. You state the major points of a larger body of information in your own words. Reread the previous paragraph. Underline the important points. Now summarize the paragraph in one or two sentences.

____

____

There is an urgent need to search for plants that can be used as medicines. The tropical rain forests contain two-thirds of the world's plant species. Many of these plant species have not been identified. But the rain forests are disappearing at a rapid rate. Useful plants may become extinct before they can be identified and studied.

One scientist is using a short cut to find useful plants. Dr. Mark Plotkin has been working with native shamans, or medicine men, in the Amazon rain forest. These men have a knowledge of plants that dates back thousands of years. They use about three hundred different plants as medicines. Dr. Plotkin will have to work quickly, though. The younger natives are more interested in modern ways. As a result, the shamans' knowledge is dying out. The plants they use may soon be gone, too.

Check your answer on page 217.

Thinking About the Article

Match each function with the part of the plant that carries it out. Write the letter of the function in the blank.

	Plant Part		Function
______	1. leaf	a.	transports water to leaves
______	2. stem	b.	outer part of the stem of a woody plant
______	3. bark	c.	area of growth
______	4. flower	d.	absorbs water and minerals from soil
______	5. bud	e.	makes food for the plant
______	6. root	f.	reproductive organ of the plant

Write your answers in the space provided.

7. Review the questions you wrote on page 56. Did the article answer your questions? If you said *yes*, write the answers. If your questions were not answered, write three things you learned from this article.

____

____

____

____

____

8. Describe two ways that plants work as medicines.

____

____

____

9. According to the table on page 57, how are reserpine and valepotriates alike?

____

____

Check your answers on page 217.

Circle the number of the best answer.

10. Which of the following plant products is used to relax muscles during surgery?

    (1) quinine

    (2) menthol

    (3) curare

    (4) digitalis

    (5) taxol

11. According to the article, how much of the world's plant species are found in tropical rain forests?

    (1) one-tenth

    (2) one-fourth

    (3) one-third

    (4) one-half

    (5) two-thirds

12. Why are drugs tested on animals before they are made available to the public?

    (1) Drugs cost a lot to produce.

    (2) Scientists need to look for a way to make a synthetic copy of the natural product.

    (3) People volunteer to take new drugs.

    (4) The effectiveness and safety of new drugs must be known.

    (5) Scientists must identify the plants before testing them.

Write your answer in the space provided.

13. Have you or someone you know ever used any of the medicines listed on page 57? If so, write why the medicine was used. If not, describe some other medicine you have taken.

    ____

    ____

    ____

    ____

Check your answers on page 217.

Animals

Setting the Stage

Wolves are meat-eaters. Most wolves live and hunt in groups that have a social order. By hunting in a group, wolves can catch animals much larger than themselves.

Past: What you already know

You may already know something about wolves. Write three things you already know.

1. ______________________________

2. ______________________________

3. ______________________________

Present: What you learn by previewing

Write the headings from the article on pages 63–65 below.

The Wolf

4. ______________________________

5. ______________________________

6. ______________________________

7. ______________________________

What does the diagram on page 64 show?

8. ______________________________

Future: Questions to answer

Write three questions you expect this article to answer.

9. ______________________________

10. ______________________________

11. ______________________________

 Check your answers on page 217.

The Wolf

As you read each section, circle the words you don't know. Look up the meanings.

Imagine camping in a Canadian forest. The silence of the evening is broken by the howl of a single wolf. Soon other wolves begin to howl. Although most people find this sound frightening, it is one way that wolves communicate with one another. Wolves live in groups called **packs**. Each pack of wolves has its own territory. Wolves howl to call the pack back together after a hunt. They also howl to warn other wolves to stay out of their territory.

The Hunter and the Hunted

The wolf is a hunter, or **predator**. The wolf's **prey** are the animals it hunts to eat. Moose, elk, deer, caribou, and ranchers' sheep and cattle are prey to wolves. These animals are larger than wolves. Hunting in packs makes it possible for wolves to go after such large prey. But the prey have good defenses against wolves.

Moose sometimes fight back when attacked. An adult moose is about eight to ten times heavier than a wolf. A well-aimed kick could break a wolf's ribs or leg. A moose can also run fast. Many moose escape simply by running away.

Elk, deer, and caribou also run away when attacked. They live in large groups called **herds**. These animals are very alert and quickly flee if they sense that wolves are nearby.

Wolves live and hunt together in a pack.

Wolves often attack an old, sick, or injured animal lagging behind the herd. By preying on weak animals, wolves help strengthen the herd. The healthy animals survive. Wolves also keep herds from getting too large, which can result in many animals starving to death.

Wolves have also been known to attack cattle and sheep. Cattle and sheep cannot outrun a pack of wolves. This makes them easy prey. Most ranchers do not like wolves. They believe wolves are a threat to their livestock and their income. Some ranchers shoot or poison wolves that are seen near their livestock.

The Wolf's Niche

Wolves are the only predators of large animals such as moose and caribou in North America. The role an animal plays in its environment is called its **niche**. Part of the role of the animal is its position in the food chain. A **food chain** shows how energy is transferred from one organism to another in the form of food.

The main source of energy in a food chain is the sun. Plants use the sun's energy to produce food. Deer are **herbivores**, or animals that eat plants. Wolves are **carnivores**, or animals that eat other animals. The word *carnivore* means "meat-eater."

This Food Chain Consists of the Sun, Grasses, Deer, and Wolves.

Applying Your Skills and Strategies

Classifying. Grouping things that are similar helps us understand how nature works. Putting things into groups is called *classifying*. In the paragraph above, the author is classifying animals according to what they eat. For example, herbivores, such as deer, eat only plants. Write the name of another animal that is a herbivore.

____

Write the name of an animal, other than the wolf, that is a carnivore.

____

The Behavior of Wolves

In a wolf pack there is a social order. In this order, there are different ranks, or levels. The dominant, or highest-ranking, wolf is the leader.

Check your answers on page 217.

Sometimes a wolf challenges the dominant wolf. The dominant wolf may growl and show its teeth. That is usually enough to send the would-be challenger on its way.

If the wolves do have a fight, it does not last long. The loser shows that it gives up by rolling over on its back. At this point, the winner could hurt or even kill the loser. But that does not happen. The winner walks away with its tail in the air and its head held high. The loser slinks away with its tail between its legs.

These two wolves have different ranks in the pack.

Applying Your Skills and Strategies

Understanding a Photo. You can learn many things from photos. The caption tells you what the photo is about. Underline the caption of the photo above. Look carefully at the photo. Compare the postures of the two wolves.

____

____

People and Wolves

Wolves have long been feared and despised. Since the late 1800s, people have killed more than 350,000 wolves in the United States. Wolves are now an endangered species in every state except Minnesota and Alaska. Conservationists want to reintroduce wolves into large wilderness areas, such as Yellowstone National Park. Recent opinion polls show that the majority of people surveyed support this idea. People are recognizing the importance and value of wolves in the food chain.

Check your answer on page 217.

Thinking About the Article

Fill in the blank with the word or words that best complete each statement.

1. Wolves live and hunt in groups called ____________.
2. Animals that hunt other animals for food are called

   ____________________.
3. The victim of the hunt is called the ____________.
4. The role an animal plays in its environment is called its

   ____________.
5. A ____________________ shows how energy is transferred from one organism to another.
6. An animal that eats only plants is called a ____________________.
7. An animal that eats only animals is called a ____________________.

Write your answers in the space provided.

8. Review the questions you wrote on page 62. Did the article answer your questions? If you said *yes*, write the answers. If your questions were not answered, write three things you learned from this article.

____

____

____

____

____

9. Look at the photo on page 63. Describe two things that you see.

____

____

____

____

Check your answers on pages 217–218.

Circle the number of the best answer.

10. What is the main source of energy in a food chain?

    (1) sun

    (2) plants

    (3) sugar

    (4) herbivores

    (5) carnivores

11. In North America, the wolf is the only

    (1) predator of large animals.

    (2) carnivore.

    (3) herbivore.

    (4) type of dog.

    (5) animal with a niche.

12. In Africa, gazelles feed on grasses and shrubs. Lions hunt gazelles. In this food chain, which animals are the herbivores?

    (1) sun

    (2) grasses and shrubs

    (3) gazelles

    (4) lions

    (5) There are no herbivores in this food chain.

Write your answers in the space provided.

13. Wolves have sharp teeth and claws. How are these things useful to wolves?

____

____

14. Do you think that wolves should be protected and reintroduced into large wilderness areas in the United States? Explain your answer.

____

____

____

Check your answers on page 218.

Life Cycles

Setting the Stage

Gypsy moths are leaf-eating insects that have damaged millions of acres of trees. Although people have tried many methods of wiping out these pests, gypsy moths continue to spread in the United States.

Past: What you already know

You may already know something about moths or other insect pests. Write three things you already know.

1. ______________________________

2. ______________________________

3. ______________________________

Present: What you learn by previewing

Write the headings from the article on pages 69–71 below.

Fighting the Gypsy Moth

4. ______________________________

5. ______________________________

6. ______________________________

What is the title of the diagram on page 70?

7. ______________________________

Future: Questions to answer

Write three questions you expect this article to answer.

8. ______________________________

9. ______________________________

10. ______________________________

Check your answers on page 218.

Fighting the Gypsy Moth

As you read each section, circle the words you don't know. Look up the meanings.

You probably take the trees in your neighborhood for granted. Yet in the Northeast, gypsy moths have stripped the leaves off millions of oak, birch, aspen, gum, and other trees. People in the northeastern United States have been fighting the gypsy moth for a hundred years.

The Spread of the Gypsy Moth

A Frenchman brought gypsy moth eggs from Europe to Massachusetts in the 1860s. He hoped to breed the moths with American silk-producing moths. Unfortunately, several moths escaped from his house. Fifty years later, the gypsy moth had spread throughout the Northeast. Today many scientists, government agencies, and private citizens are fighting the gypsy moth as far south as Virginia and as far west as Michigan. The moths have been seen even in California and Oregon.

Gypsy moth caterpillars do a great deal of damage. They will eat every leaf on trees they like, such as oak. However, they skip other types of trees, such as spruce. Healthy trees can survive a gypsy moth attack. The leaves usually grow back the next year. But trees weakened by drought or disease may die after a year or two.

In addition to damaging trees, gypsy moths make a mess. They leave half-eaten leaves everywhere. Their droppings ruin the finish on cars. Gypsy moth caterpillars make their way onto porches, screens, and windows. When the caterpillars die, the smell of decay is terrible.

Red Oak tree damaged by gypsy moths

The Life Cycle of the Gypsy Moth

To find ways to get rid of gypsy moths, scientists study the insects' life cycle. A **life cycle** is the series of changes an animal goes through in its life. There are four stages in the life cycle of a gypsy moth.

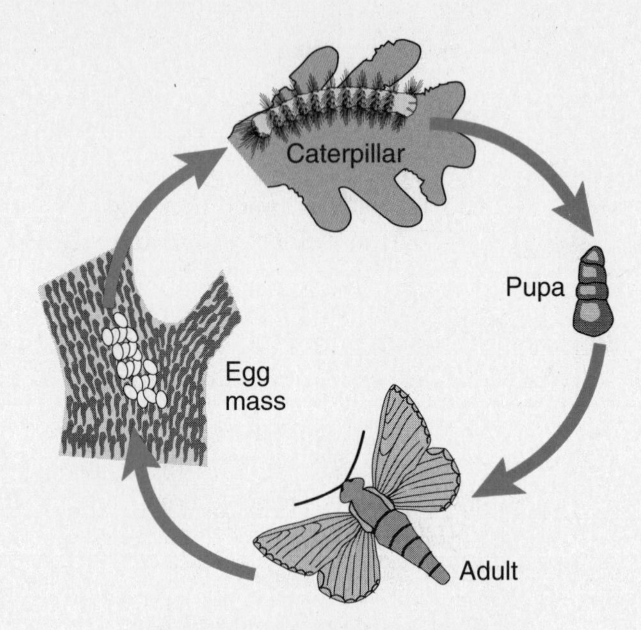

The Gypsy Moth's Life Cycle

1. **Egg.** Gypsy moth eggs are laid during the summer. Each female lays from 75 to 1,000 eggs in a clump called an **egg mass**. Egg masses are tan and slightly fuzzy.
2. **Caterpillar.** Gypsy moth caterpillars climb up trees to find leaves to eat. The caterpillars grow from a fraction of an inch to more than two inches long.
3. **Pupa.** During the pupa stage, the caterpillar encloses itself in a case for about two weeks. The body changes into an adult.
4. **Adult.** When the pupa case breaks open, the adult moth comes out. Adult moths do not eat, so they do not live long. The female gives off a smell that attracts males. After mating, the male dies. The female lives long enough to lay eggs, and then she dies.

Applying Your Skills and Strategies

Sequence. Sequence means the order in which things happen. Sequence tells what comes first, second, third, and so on. Often, things that happen in sequence are numbered, as in the stages of the life cycle on this page. List the four stages of the gypsy moth's life, beginning with the egg. What would the next stage be if the cycle were to continue?

___

___

Check your answer on page 218.

A life cycle is often shown in the circular style used on page 70. However, it can also be shown in a timeline, such as the one below.

Life Cycle of a Gypsy Moth

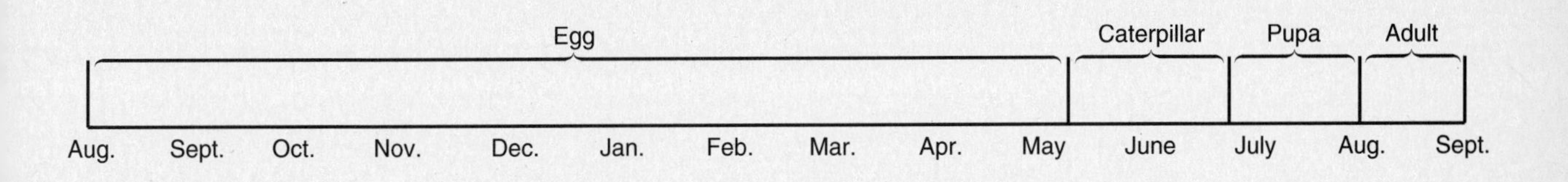

Applying Your Skills and Strategies

Reading a Timeline. A timeline is another way to show a sequence. When you see a timeline, look at its title first. That will tell you what the timeline is about. Underline the title of the timeline above.

Next, look at the labels. The labels along the bottom are months. The labels at the top show the stages of a gypsy moth's life. What does the timeline tell you about these stages?

____

____

Controlling the Gypsy Moth

It is when the gypsy moth is in the caterpillar stage that it does all the damage to trees. The caterpillars can be killed by spraying the trees. Some sprays are chemicals that can harm other animals and the environment. Other sprays contain bacteria that kill only the moths. Caterpillars can be caught by wrapping sticky tape around trees. The caterpillars can also be killed by destroying egg masses before they hatch. Egg masses can be found on trees, buildings, fences, and outdoor furniture. The egg masses can be scraped into a bucket of kerosene, bleach, or ammonia.

Pupa cases can be removed from trees and crushed. Adult males can be caught in traps. The traps contain a bait that smells like a female gypsy moth. The male flies into the trap and dies there.

Some birds prey on the moth. Birds that eat the moths can be attracted into an infested area. People can put out food, water, and nesting materials to encourage birds to stay. Another natural control method uses parasites. A **parasite** is an organism that lives on or in another organism and harms it. Many types of parasites have been released in areas with large numbers of gypsy moths.

Diseases can lower the number of moths. One year there was a very rainy spring, and many moths died of a fungus disease. Perhaps one day scientists will be able to use the fungus to kill gypsy moths.

Check your answer on page 218.

Thinking About the Article

Fill in the blank with the word or words that best complete each statement.

1. The __________ is the first stage of the gypsy moth's life cycle.
2. The caterpillar is enclosed in a case during the __________ stage.
3. A ______________ is an organism that lives on or in another organism and harms it.

Write your answers in the space provided.

4. Review the questions you wrote on page 68. Did the article answer your questions? If you said *yes,* write the answers. If your questions were not answered, write three things you learned from this article.

____

____

____

____

____

5. According to the timeline on page 71, which stage of the gypsy moth's life cycle lasts the longest?

____

____

Match the stage of the gypsy moth's life cycle with the control method used during that stage. Write the letter of the stage in the blank.

	Control Method	Stage
______	6. spraying insecticide	a. egg
______	7. soaking in ammonia	b. caterpillar
______	8. scent traps	c. pupa
______	9. crushing	d. adult

 Check your answers on page 218.

Circle the number of the best answer.

10. During which stage of its life does the gypsy moth do the most damage?
    (1) egg
    (2) caterpillar
    (3) pupa
    (4) parasite
    (5) adult

11. Which control method would cause the most damage to organisms other than the gypsy moth?
    (1) chemical sprays
    (2) bacterial sprays
    (3) soaking egg masses in bleach
    (4) scent traps
    (5) crushing the pupa

12. According to the article, adult moths do not eat. The purpose of catching adult males with scent traps is to
    (1) prevent males from flying to new places.
    (2) keep males from mating with females, who then lay eggs.
    (3) save the leaves of oak, birch, aspen, and gum trees.
    (4) avoid having to soak the moths in kerosene.
    (5) use males in scientific research.

Write your answers in the space provided.

13. During the month of February, which control method can people use to kill gypsy moths?

____

14. Describe an experience you have had with gypsy moths or other insect pests. What did you do to control them?

____

____

____

____

Check your answers on pages 218–219.

The Environment

Setting the Stage

Each year billions of tons of waste are produced in the United States. Most of this waste winds up in landfills. With little air or water to help break it down, the garbage is preserved for years.

Past: What you already know

You may already know something about landfills or garbage. Write three things you already know.

1. ______________________________

2. ______________________________

3. ______________________________

Present: What you learn by previewing

Write the headings from the article on pages 75–77 below.

Where Does the Garbage Go?

4. ______________________________

5. ______________________________

6. ______________________________

7. ______________________________

What does the graph on page 76 show?

8. ______________________________

Future: Questions to answer

Write three questions you expect this article to answer.

9. ______________________________

10. ______________________________

11. ______________________________

 Check your answers on page 219.

Where Does the Garbage Go?

As you read each section, circle the words you don't know. Look up the meanings.

Each day the average American tosses out almost four pounds of **municipal solid waste**. Most people call it garbage. Whatever you call it, each person throws away over half a ton of garbage per year, or 160 million tons per year total. The Environmental Protection Agency estimates that by the year 2000, we will each throw away 4.4 pounds every day. That's 216 million tons of garbage per year. That figure doesn't include the billions of tons produced each year by businesses and farms.

What We Throw Away

Garbage contains two kinds of materials. **Organic materials** come from things that were once alive. Paper, food waste, yard waste, and wood are all organic. **Inorganic materials** were never alive. They include plastics and metals.

Most garbage is organic. Paper makes up almost half of all garbage. Yard waste makes up almost a fifth. The rest is glass, metals, plastics, food, and other materials, including disposable diapers.

Where does all this garbage go? About 13 percent is **recycled**, or processed to be used again. Newspapers, aluminum cans, glass jars, and some plastic containers can be recycled. Fourteen percent of garbage is **incinerated**, or burned. Dumping garbage in the ocean is no longer allowed. So most of our garbage winds up in **landfills**, which are modern versions of the town dump.

A crane unloading garbage into a landfill

You can learn a lot about people by looking at what they throw away. This bar graph shows what has been going to the landfill or incinerator after recycled materials are removed.

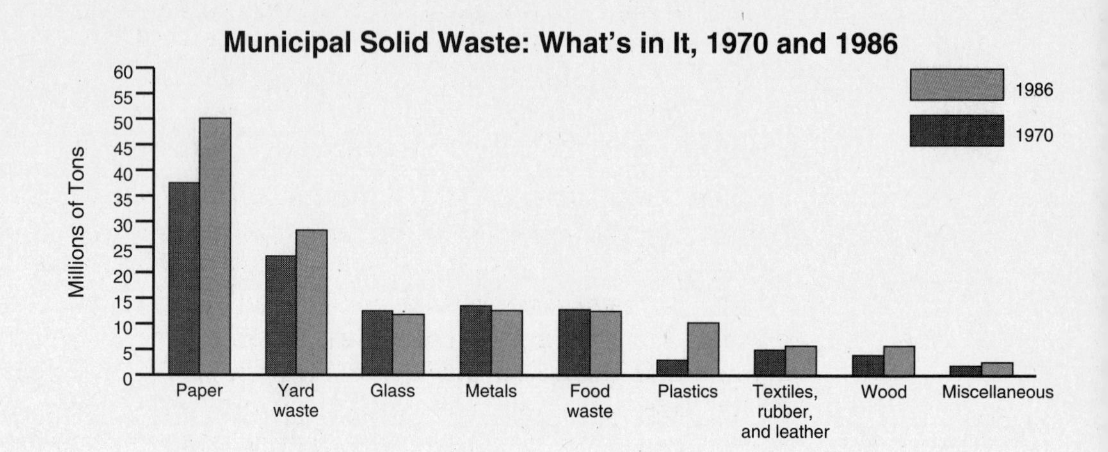

Applying Your Skills and Strategies

Reading a Bar Graph. Bar graphs are used to compare sets of information. The graph above compares the amount of each type of garbage in two different years. The height of the bars shows you at a glance whether an item increased or decreased over time. To read a bar graph, first read the title. Circle the title in the graph above. Then read the words along the vertical line to see what measurement is being used. What is that measurement?

____

Read the words along the horizontal line. What do these words tell?

____

Look at the key to the graph. The key tells what symbols are used. The colored bars show 1986 figures. What do the black bars show?

____

Which garbage items decreased from 1970 to 1986?

____

What Happens in a Landfill?

Inorganic materials break down very slowly, if at all. They may remain for hundreds of years. That is why these materials should be recycled, not dumped. Organic garbage **degrades**, or breaks down, through the action of living things. Living things that break down organic materials are called **decomposers**. Many product labels claim to be "biodegradable." But these labels can be misleading.

 Check your answers on page 219.

Many people think that organic garbage degrades in a landfill. But this is not true. Most decomposer bacteria are **aerobic**, which means they need oxygen to live. In a landfill, there is no oxygen below eight to ten feet. Since the aerobic bacteria cannot survive, the garbage remains. The University of Arizona's Garbage Project recently dug up an old landfill. They found hot dogs, carrots, and corn cobs—all intact. A newspaper they found from 1952 was still readable.

What Happens in a Compost Pile?

A few places are starting to compost garbage instead of using landfills. In **composting**, organic material is broken down by aerobic bacteria, fungi, insects, and worms. Unlike landfills, compost piles have oxygen to keep bacteria alive. The oxygen is provided by turning the garbage periodically. After a while, the garbage turns into a soil-like mixture called compost. This can be added to soil for gardening or farming. People have been composting in their yards for years. Large-scale commercial composting is fairly new. So far only ten states have composting plants.

Applying Your Skills and Strategies

Applying Knowledge to Other Contexts. When knowledge is put to use in a new situation, it can be used to solve problems. Scientists are always trying to apply ideas to new situations. In the paragraph above, the author describes an old idea, composting. How is composting being applied to a new situation?

___

___

What You Can Do

Labels such as this show that a product can be recycled.

Some cities encourage recycling by charging people for each can of garbage that is picked up. In Seattle, for example, the more garbage you make, the more you pay to have it taken away. There are many things you can do to reduce the amount of garbage you throw away. You may not save money unless you live in a city like Seattle, but you will help save the environment.

- Take part in a recycling program. Separate items for recycling, such as tin and aluminum cans, glass, plastic, and newspapers.
- Recycle products in your home. Wash and reuse plastic bags and jars.
- Make less trash in the first place. Look for products that can be used more than once. Buy items that have recyclable packaging. Pack groceries in canvas bags instead of plastic or paper bags.
- If you have a yard, start a compost pile. Even if you compost only yard waste, you can make a big difference.

Check your answer on page 219.

Thinking About the Article

Fill in the blank with the word or words that best complete each statement.

1. When waste is processed to be used again, it is being

   ________________.

2. Some garbage is burned, or ________________________.

3. Most garbage winds up in dumps, or ____________________.

4. ________________________ materials are made from substances that were never alive.

5. Paper and food waste are examples of ________________ materials.

6. ________________________ are organisms that break down organic garbage.

Write your answers in the space provided.

7. Review the questions you wrote on page 74. Did the article answer your questions? If you said *yes,* write the answers. If your questions were not answered, write three things you learned from this article.

____

____

____

____

____

8. Refer to the bar graph on page 76. Which item accounted for the most garbage in 1970?

____

9. Why does most organic garbage not degrade in a landfill?

____

____

____

Check your answers on page 219.

Circle the number of the best answer.

10. What portion of municipal solid waste is recycled?

    (1) 5 percent

    (2) 13 percent

    (3) 20 percent

    (4) 50 percent

    (5) 100 percent

11. According to the bar graph on page 76, which of the following wastes tripled in weight between 1970 and 1986?

    (1) paper

    (2) yard waste

    (3) glass

    (4) plastics

    (5) wood

12. Organic garbage degrades better in a compost pile than in a landfill because only compost piles

    (1) contain plastics.

    (2) are turned to maintain oxygen levels.

    (3) contain food wastes.

    (4) contain bacteria.

    (5) are used for gardening.

Write your answers in the space provided.

13. Suppose a new brand of bread or other food had recyclable packaging while your favorite brand didn't. Would you change to the new brand? Explain your answer.

____

____

14. What items do you recycle regularly? If you do not recycle now, what items would be easiest for you to recycle?

____

____

Check your answers on page 219.

Ecosystems

Setting the Stage

Tropical rain forests are warm, humid areas. They have a wide variety of plants and animals. Although rain forests cover only a small part of Earth, they benefit the whole world. The rain forests are disappearing because of farming, logging, and ranching.

Past: What you already know

You may already know something about rain forests or other types of forests. Write three things you already know.

1. ______________________

2. ______________________

3. ______________________

Present: What you learn by previewing

Write the headings from the article on pages 81–83 below.

Tropical Rain Forests

4. ______________________

5. ______________________

6. ______________________

7. ______________________

What does the diagram on page 82 show?

8. ______________________

Future: Questions to answer

Write two questions you expect this article to answer.

9. ______________________

10. ______________________

Check your answers on page 219.

Tropical Rain Forests

As you read each section, circle the words you don't know. Look up the meanings.

Do you like rice cereal with bananas, coffee with sugar? Did you know that the foods in this breakfast first came from tropical rain forests? Tropical rain forests are the source of many foods that are now grown commercially.

Take a look around you. Offices, libraries, and department stores may have furniture made of mahogany or teak. These woods grow in tropical rain forests. Check out your medicine cabinet. One out of every four drugs has ingredients that first came from tropical rain forest plants.

What Is a Tropical Rain Forest?

A tropical rain forest is a large ecosystem. An **ecosystem** is an area in which living and nonliving things interact. Tropical rain forests are found near the equator. In these areas, there is year-round sun, warmth, and rain. The temperature may vary more from day to night than from season to season.

The conditions in tropical rain forests support a wealth of plant and animal life. More than half of all types of plants and animals live in tropical rain forests. The variety of life is tremendous. For example, just one square mile of rain forest in Peru has 1,450 types of butterflies. Together, the United States and Canada have only 730!

A tropical rain forest

The Global Effects of the Rain Forests

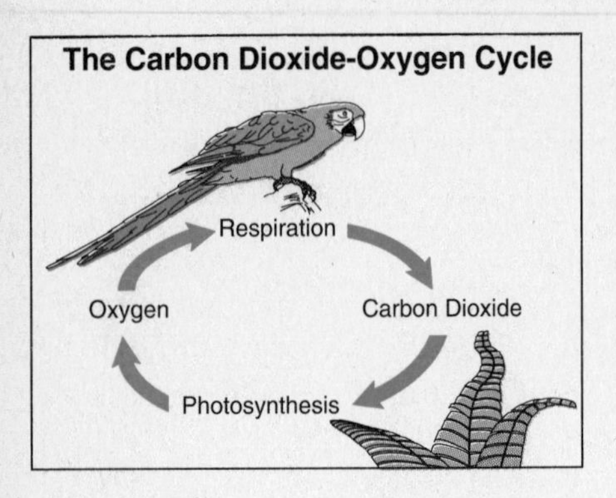

The plants of the rain forests absorb carbon dioxide from the atmosphere. Through the process of **photosynthesis**, plants use carbon dioxide, water, and energy from sunlight to make food. This process gives off oxygen. Oxygen is used by animals for respiration. **Respiration** is the process by which living things get energy from food. This process gives off carbon dioxide. Carbon dioxide is also given off when fuels are burned and organic material, such as plants and animals, decompose. The carbon dioxide is then used by plants, completing the **carbon dioxide–oxygen cycle**.

Applying Your Skills and Strategies

Recognizing Cause and Effect. Situations in which one thing makes another thing happen have a cause-and-effect relationship. For example, the sun shines on a pole (cause). The pole casts a shadow (effect). Science is full of cause-and-effect relationships. Watch for words such as *cause, effect, because, result, leads to, due to, therefore, thus,* and *so*. These words often signal cause-and-effect relationships.

In the previous diagram and paragraph, you learned about the carbon dioxide–oxygen cycle. If more forests were planted, what would be the effect on the amount of carbon dioxide in the air?

____

____

The amount of carbon dioxide in the air affects Earth's temperature. Carbon dioxide absorbs heat. The more carbon dioxide in the atmosphere, the warmer Earth gets. This warming is known as the **greenhouse effect**. By taking carbon dioxide from the air, the plants of the rain forests help control Earth's temperature.

Destruction of the Rain Forests

Tropical rain forests are being destroyed at a rate of one hundred acres per minute. That's almost one football field each second. Destruction of the rain forests is mainly caused by farmers, ranchers, and loggers. Many tropical countries don't have enough farmland to feed their people. So farmers clear forests by cutting down trees and burning them. After a few years the soil loses its nutrients. Then the farmers move on to clear and burn new places. Ranchers cut down trees to make room for their cattle. Loggers also destroy the forests by cutting down trees for the wood.

This destruction affects the carbon dioxide–oxygen cycle in two ways. First, many trees are burned to clear the land. This burning adds tons of carbon dioxide to the air. Second, destroying trees leaves fewer plants to take carbon dioxide from the air. Scientists think that more carbon dioxide in the air will cause Earth to warm up.

Check your answer on page 219.

Tropical rain forests cover 6 to 7 percent of Earth's land area. They are found along the equator, as shown in the map below. Large areas of the rain forests have already been destroyed.

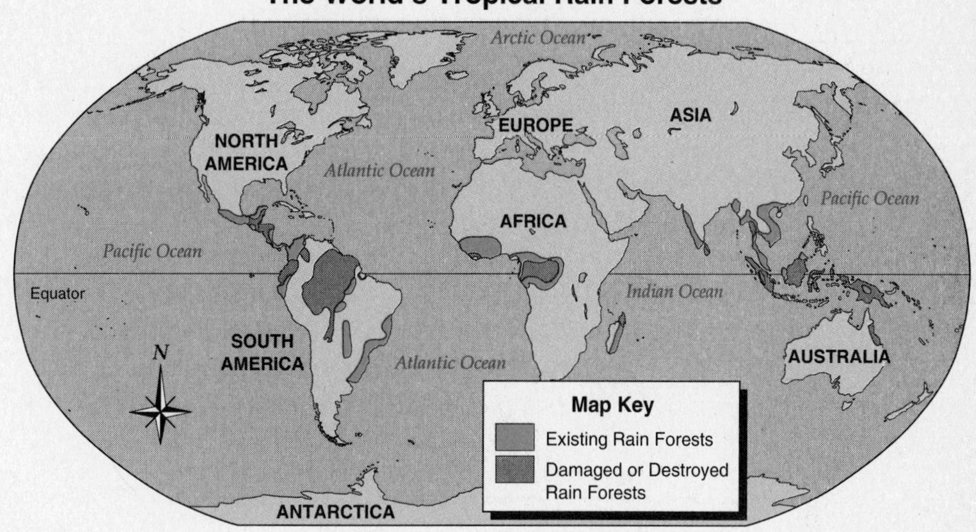

Applying Your Skills and Strategies

Reading a Map. To study a map, first look at the title. It tells you what the map shows. Circle the title of the map above. Then look for the *map key*. The map key tells you how information is shown on the map. How are tropical rain forests shown?

____

Look at the *compass rose*. It shows the directions north, south, east, and west. Where are most tropical rain forests in relation to the equator?

____

Stopping the Destruction

People in tropical countries do not destroy the forests on purpose. In these countries, the forests are a source of land, fuel, food, and cash. It's not possible to keep farmers, ranchers, and loggers out of the rain forests. Many countries are now trying to save some areas in the forests by making them off-limits. New methods of farming and logging can use the forests without destroying them.

Other countries can help, too. They can buy products, such as Brazil nuts, that are grown without damaging the forests. They can refuse to buy products that damage the forests. This includes beef ranched in the forests. It also includes tropical woods, such as teak.

Check your answers on page 219.

Thinking About the Article

Fill in the blank with the word or words that best complete each statement.

1. An area in which living and nonliving things interact is called an

   ____________________.

2. _____________________________ is a process by which plants use carbon dioxide, water, and energy from sunlight to make food.

3. _______________________ is a process by which living things use oxygen to get energy from food.

4. The ____ describes how carbon dioxide and oxygen circulate through the world.

Write your answers in the space provided.

5. Review the questions you wrote on page 80. Did the article answer your questions? If you said *yes,* write the answers. If your questions were not answered, write three things you learned from this article.

6. What are some foods we eat that originally came from tropical rain forests?

7. What role do animals play in the carbon dioxide–oxygen cycle?

Check your answers on page 220.

Circle the number of the best answer.

8. All of the following processes increase the amount of carbon dioxide in the air except
   (1) photosynthesis.
   (2) respiration.
   (3) burning fuels.
   (4) decomposition of organic material.
   (5) All of the above increase the amount of carbon dioxide in the air.

9. People burn more fuel now than ever before. The effect of this is that the amount of carbon dioxide in the air has
   (1) increased.
   (2) decreased.
   (3) first increased, then decreased.
   (4) first decreased, then increased.
   (5) remained the same.

10. Refer to the map on page 83. Which of the following statements is not true?
   (1) Most tropical rain forests are located near the equator.
   (2) Australia has tropical rain forests.
   (3) All tropical rain forests are located south of the equator.
   (4) There are no tropical rain forests in North America.
   (5) South America has a large area of tropical rain forests.

Write your answers in the space provided.

11. List two ways that tropical rain forests are important.

____

____

12. What can you do to help prevent the destruction of tropical rain forests?

____

____

____

Evolution

Setting the Stage

People use poisons called pesticides to get rid of unwanted animals, such as rats. Sometimes rats become resistant to a poison. The poison no longer works because the rat population has changed.

Past: What you already know

You may already know something about rats or pesticides. Write two things you already know.

1. ______________________________

2. ______________________________

Present: What you learn by previewing

Write the headings from the article on pages 87–89 below.

Adaptable Rats

3. ______________________________

4. ______________________________

5. ______________________________

6. ______________________________

7. ______________________________

What do the drawings on page 89 show?

8. ______________________________

____

Future: Questions to answer

Write two questions you expect this article to answer.

9. ______________________________

10. ______________________________

 Check your answers on page 220.

As you read each section, circle the words you don't know. Look up the meanings.

Adaptable Rats

How big do you think the average rat is? Is it the size of a cat? If you're like most people, rats loom large in your mind. They have a reputation for being dirty, sneaky, and daring. Yet rats are actually quite small. Most weigh under a pound and measure less than ten inches, not counting their tails. But rats stir up a great deal of fear and disgust in most people.

The City Rat and the Country Rat

Rats need a large amount of food. They eat a quarter of their weight each day. Rats found that it's easier to eat our food than to hunt for food in the wild. Rats live among people in the city, suburbs, and country. In the city, they live off garbage. In the suburbs, bird feeders and dog food bowls provide easy meals. In the country, barns become storehouses of rat food. Rats come out at night, so people rarely see them. What people do see is the damage rats cause. They destroy property, contaminate food, and spread disease.

Rat Control

People wage a constant war against rats. Poisons and traps have been used with varying degrees of success. Keeping the rat population under control requires new weapons all the time.

A rat eating corn in storage

In one North Carolina case, rats took over a farmer's barn after the harvest. He called an exterminator for help. The exterminator put out food covered with *Warfarin*, a common rat poison. Warfarin had worked before. So the farmer thought his rat problem was over. But this time, the rats ate both the crops and the poisoned food with great appetite. They didn't even get sick.

Applying Your Skills and Strategies

Making Inferences. An *inference* is a fact or idea that follows logically from what has been said. Readers make inferences from what they read all the time. Review the first paragraph on this page. From what the author says, you can infer that something went wrong. Write one other thing you can infer from this paragraph.

____

____

Coloring is an example of protective adaptation.

Adapting to Warfarin

Why didn't the rats die from Warfarin? Rats are becoming resistant to Warfarin. That means they can eat it and not be harmed. The rats in North Carolina had adapted to Warfarin. A trait that makes a plant or an animal better able to live in its environment is called an **adaptation**. For example, an arctic fox's white fur helps it blend into the background of snow. Its enemies have trouble seeing it, so the fox has a better chance of surviving. The rat's ability to eat Warfarin without ill effects helps it survive.

Many adaptations occur through mutation. A **mutation** is a change in a gene. Mutations happen by chance and most are harmful. But sometimes a mutation causes a trait that helps an organism survive.

Evolving Rats

There are always variations in any population due to mutation. A few rats in a population may be resistant to Warfarin. A rat with an adaptation, such as resistance to Warfarin, is more likely to survive and reproduce. A rat without this adaptation is more likely to die before it reproduces.

Charles Darwin, a nineteenth-century English scientist, called this process natural selection. **Natural selection** means that the organisms best suited to their environments are most likely to survive. Natural selection is sometimes called *survival of the fittest*. What *fit* is depends on the situation. In this case, Warfarin caused some rats to die and others to survive. However, survival of an individual is not enough. The individual must reproduce to pass on the adaptations.

Check your answer on page 220.

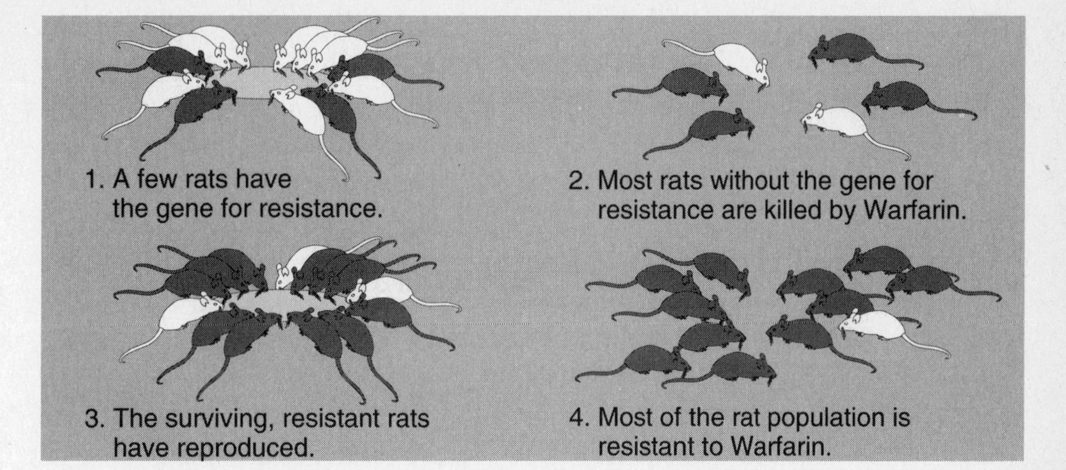

The resistance to Warfarin evolves through natural selection. The dark rats are resistant.

The make-up of the population changes slowly. Organisms with adaptations reproduce more than others. The useful traits will appear in more of the offspring. Over time the adaptation may be found in most of the population. This gradual change in a species over time is called **evolution**.

Adaptations can spread fast in species that breed quickly. Rats are fast breeders. A healthy female may give birth to sixty or seventy rats per year. The young rats can reproduce when they are only a few months old. It doesn't take long for an adaptation like resistance to Warfarin to spread through a population of rats.

Applying Your Skills and Strategies

Applying Knowledge to Other Contexts. Using the knowledge you gain in one situation and applying it to another can give you new insights about the world around you. For example, you just learned how rats developed resistance to Warfarin. How does natural selection apply to the following situation?

Insecticides are used to kill insects. But if you use them too often, they no longer work.

____

____

Changing Tactics

When rats become resistant to one poison, exterminators must try others. When Warfarin is removed from the environment, resistance to it will no longer be a useful trait. Rats without this adaptation will have as much chance of reproducing as rats with it. Over many generations the resistance may become less common. At that point, Warfarin can be used again. But if there are any resistant rats still around, it won't work for long.

Check your answer on page 220.

Thinking About the Article

Fill in the blank with the word or words that best complete each statement.

1. A trait that makes a plant or animal better able to live in its environment is called an ________________.
2. A ________________ is a change in a gene.
3. ________________________ means that organisms best suited to their environments are most likely to survive and reproduce.
4. The gradual change in a species over time is called ________________.

Write your answers in the space provided.

5. Review the questions you wrote on page 86. Did the article answer your questions? If you said *yes*, write the answers. If your questions were not answered, write three things you learned from this article.

____

____

____

____

6. Why do people rarely see rats?

____

7. Why did Warfarin stop working as a rat poison?

____

Circle the number of the best answer.

8. Which characteristic is an adaptation that would help an arctic fox survive?

   (1) blue eyes
   (2) white fur
   (3) short fur
   (4) slow pace
   (5) poor ability to smell

Check your answers on page 220.

9. *Survival of the fittest* means nothing to evolution unless the fittest organisms

   (1) mutate.

   (2) get sick.

   (3) adapt.

   (4) reproduce.

   (5) die.

10. Resistance to Warfarin stops being a favorable adaptation for rats when

    (1) all rats die out.

    (2) Warfarin is no longer present in the environment.

    (3) Warfarin is used to control rats.

    (4) there is a decrease in the number of baby rats.

    (5) rats move to other areas.

Write your answers in the space provided.

11. Why do favorable adaptations spread quickly in populations that reproduce quickly?

    ____

    ____

12. Cactus plants are juicy and full of water. They live in the desert, where water is scarce. These plants are covered with sharp spines. How might the spines of the cactus be a useful adaptation?

    ____

    ____

    ____

13. Describe an experience you or someone you know has had with animal, insect, or plant pests.

    ____

    ____

    ____

    ____

Check your answers on pages 220–221.

Unit 1 Review:
Life Science

How Cells Reproduce

The **nucleus** of a cell is the control center. In the nucleus are the chromosomes. They contain the genetic material, or **DNA**. This material has all the instructions the cell needs to live. When the cell reproduces, this information is passed on to the new cells.

A cell reproduces by dividing. This five-step process is called mitosis. In **mitosis**, one cell becomes two cells. These two cells can grow and divide again. Before a cell can divide, it makes a copy of the DNA. The two copies are then split up between the two new cells when the chromosomes split.

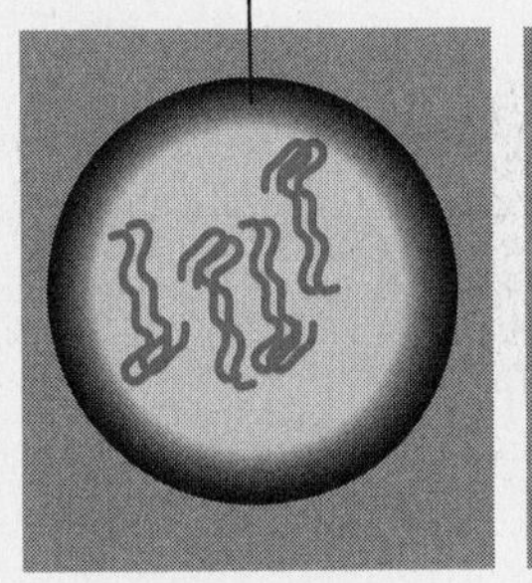

DNA doubles in parent cell

Chromosomes shorten

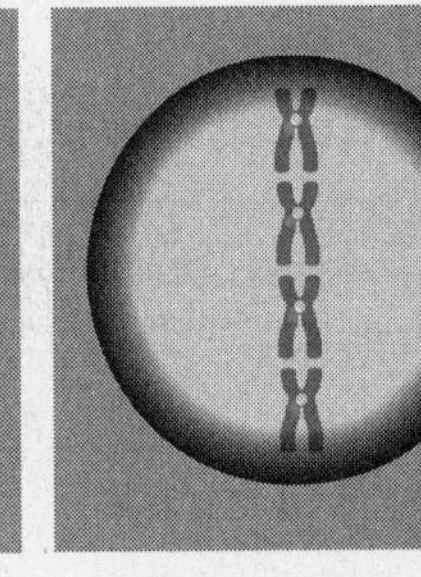

Chromosomes line up

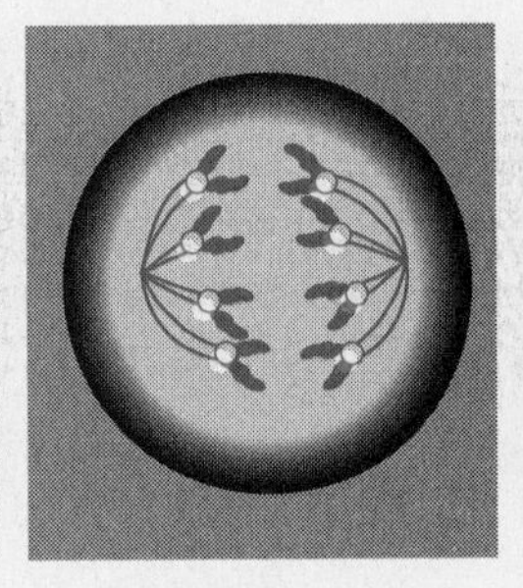

Chromosomes split

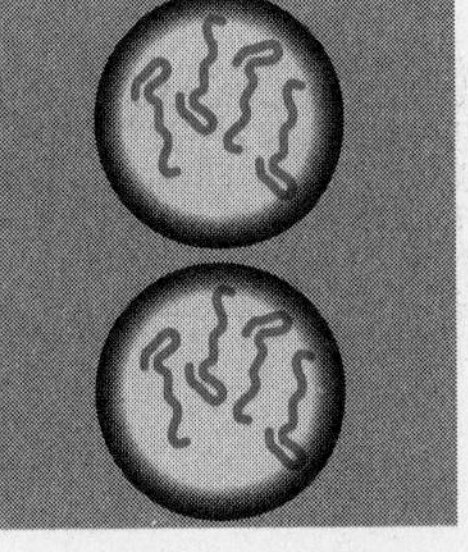

Two new cells form

Fill in the blank with the word or words that best complete each statement.

1. The genetic material is called __________.

2. Cells reproduce by division, or ______________.

Circle the number of the best answer.

3. The DNA in a cell is copied before the cell divides so that
   (1) the parent cell won't run out of DNA.
   (2) the parent cell will be able to grow.
   (3) the two new cells will have all the information they need.
   (4) the two new cells will be larger than the parent cell.
   (5) the chromosomes will be fatter and easier to see.

Go on to the next page.

The Heart

Your heart is a large, muscular pump. As you read this, your heart is pumping blood throughout your body. The blood carries oxygen to the body. Your heart is also pumping blood to your lungs, where the blood absorbs oxygen. The parts of the heart are shown below.

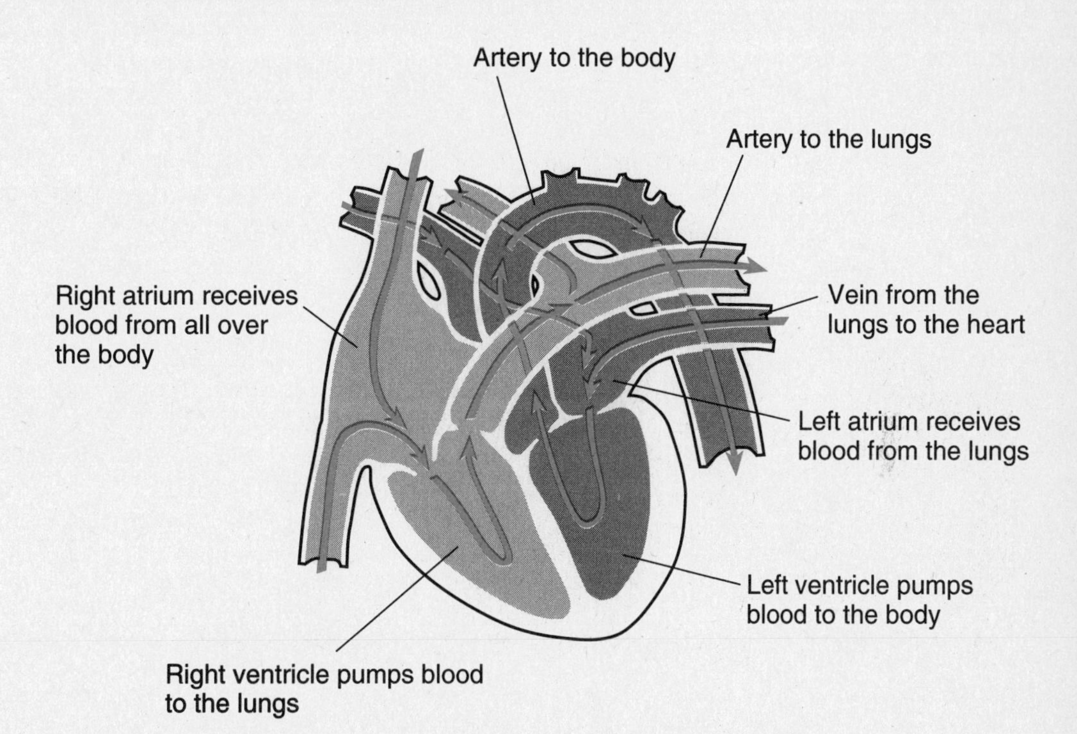

Fill in the blank with the word or words that best complete each statement.

4. If blood is flowing away from the heart, it is in an ______________.
5. Blood coming from the lungs flows through a vein to the

   ____________________ of the heart.

Circle the number of the best answer.

6. Which structure has the thickest walls?
   (1) artery
   (2) vein
   (3) atrium
   (4) ventricle
   (5) They all have walls of the same thickness.

Immunity

We are surrounded by germs. Even though bacteria and viruses are all around us, they rarely make us ill. White blood cells protect the body against germs. Any germs that enter the body are seen as foreign substances. A foreign protein is called an **antigen**. When an antigen is in the body, the white blood cells begin to make antibodies. An **antibody** is a protein the body makes to defend itself. Antibodies attack and kill invading germs.

When you are immune to a disease, you have antibodies that protect you. One way to form antibodies is to get the disease. While you are sick, your body makes antibodies, which kill the germs. After the illness is over, you still have the antibodies. So the next time you come in contact with the disease, you will be immune to it. The antibodies you developed will protect you from getting sick again.

You can also become immune to a disease without becoming ill. Scientists have found that they can teach your body to make antibodies. A **vaccine** contains weakened bacteria or viruses. These weakened germs do not make you ill. But your white blood cells recognize them and make antibodies.

Some immunities last a lifetime. You may have been vaccinated for smallpox as a child. You will probably not need to have that done again. Other vaccinations have to be repeated. You have probably had more than one tetanus shot in your life. The booster shot reminds the white blood cells to make more antibodies.

Fill in the blank with the word or words that best complete each statement.

7. Proteins that protect the body are called ________________.
8. If you have had measles, you will not get them again. This is because

   you are ____________ to measles.

Circle the number of the best answer.

9. Scientists are working on a vaccine for the AIDS virus. The first step is

   (1) making white blood cells attack the virus that causes AIDS.

   (2) weakening the virus that causes AIDS.

   (3) identifying the virus that causes AIDS.

   (4) making antibodies against the virus that causes AIDS.

   (5) killing the virus that causes AIDS.

Go on to the next page.

An Inherited Trait

Can you roll your tongue? Try it! Stick out your tongue and roll it into a tube. Some people can do this easily. Other people cannot do it at all. No matter how hard they try, they never will be able to roll their tongue. Like blue eyes or red hair, the ability to roll the tongue is an inherited trait.

Your characteristics are determined by the **genes** you inherited from your parents. Genes come in pairs. One gene from each pair came from each parent. If you can roll your tongue, it is because each parent gave you the gene for that trait. The gene for tongue rolling is **recessive**. It shows up only if there is not another gene hiding it. The gene for not rolling the tongue is dominant. A **dominant** gene can hide the recessive gene. So if you have the dominant gene, you cannot roll your tongue.

What if you have one of each type of gene? You will show the dominant form of the trait and will not be able to roll your tongue. However, the recessive gene is still there. It might be passed on to a child. In this way, recessive genes may be hidden for many generations in a family.

Fill in the blank with the word or words that best complete each statement.

10. A ____________________ gene can hide the presence of another gene.

11. The ability to roll the tongue is the result of a pair of

    ____________________ genes.

12. Characteristics are determined by the ____________ that are inherited from the parents.

Circle the number of the best answer.

13. Right-handedness is a dominant trait. If two right-handed parents have a left-handed child, it is because

    (1) neither parent has a recessive gene.

    (2) only one parent has a recessive gene.

    (3) each parent has one dominant gene and one recessive gene.

    (4) both parents have two dominant genes.

    (5) both parents have two recessive genes.

Go on to the next page.

The Nitrogen Cycle

All living things need nitrogen in some form. Nitrogen is an important part of proteins. Proteins make up much of the structure of living things. For example, muscles are made mostly of protein.

The air around you is 78 percent nitrogen. But your body cannot use it. Only a few kinds of bacteria can use nitrogen from the air. These nitrogen-fixing bacteria live in the soil. They change nitrogen into a form called **nitrates**. Plants take in the nitrates from the soil. Plants make protein, which animals get when they eat the plants. When plants and animals die, they decay. Some bacteria return the nitrogen to the soil. Other bacteria release nitrogen into the air.

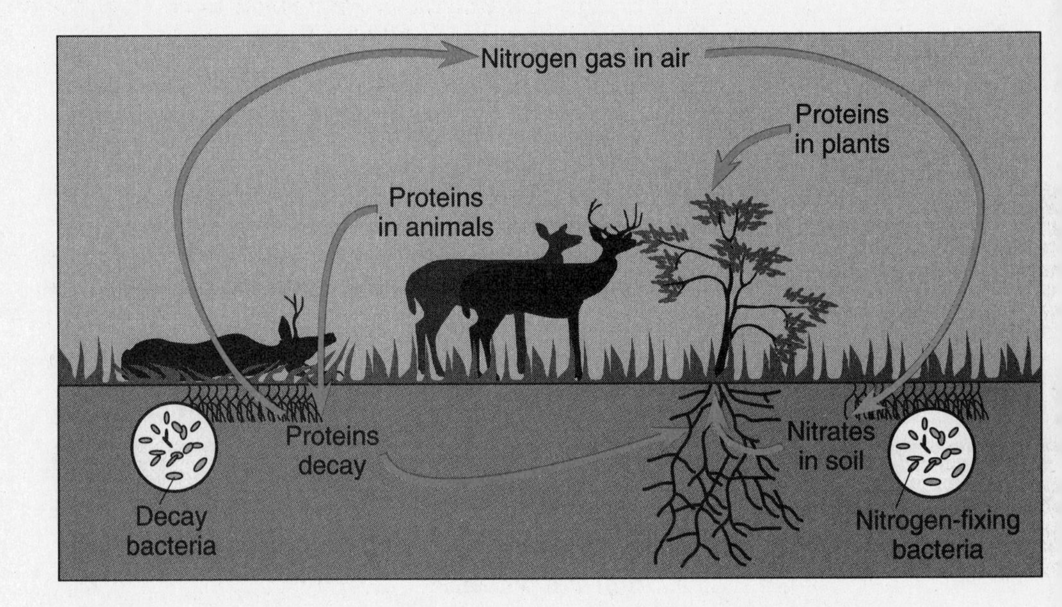

Match the organism with its source of nitrogen. Write the letter of the source in the blank at the left.

	Organism	Nitrogen source
______	14. animals	a. air
______	15. plants	b. protein
______	16. nitrogen-fixing bacteria	c. nitrates

Fill in the blank with the word or words that best complete the statement.

17. Lightning can cause a chemical reaction in which nitrogen from the air is changed into nitrates. This is like the action of the

_______________________________________.

The Ant and the Acacia

You have probably seen ants swarming all over a piece of candy dropped on the sidewalk. Cleaning up such bits of food and eating other dead animals is the role that most ants fill in nature. These ants are called scavengers.

Not all ant species are scavengers. Some species of ants live in a partnership with the acacia plant. Many partnerships in nature involve a give and take. Each partner gives the other something it needs. The need may be food, shelter, water, or protection from an enemy. In the case of the ant and the acacia tree, the trade is food and shelter for defense. A relationship in which two species help each other is called **mutualism**.

The acacia makes a sugary sweet nectar, which the ants eat. The ants also live in the thorns of the acacia plant. The ants protect the plant from animals that might eat it. If a deer starts nibbling on an acacia leaf, the ants swarm and sting it all over. The deer will think twice before eating an acacia leaf again. If any other plants start to grow near the acacia, the ants chew them down.

Fill in the blank with the word or words that best complete each statement.

18. In its relationship with the acacia plant, the ant gets ____________ and ____________.

19. In its relationship with the ant, the acacia plant gets ____________.

20. Animals that eat other dead animals are called ____________.

Circle the number of the best answer.

21. Which of the following is an example of mutualism?
    (1) A bird eats ticks that are on the back of an ox.
    (2) An ant cleans up a sidewalk by eating spilled food.
    (3) Ants carry bits of leaves back to the anthill.
    (4) A wild dog eats what is left of an antelope after lions have finished eating.
    (5) Two chimpanzees remove fleas from each other's fur.

Check your answers on page 221.

EARTH SCIENCE

Photo of Earth taken from space

Earth science is the study of Earth and its surroundings. It includes Earth's interior and its surface—both land and water. Earth science is also the study of the atmosphere and outer space. Earth science covers topics as large as the universe and as small as a stone. Studying Earth science helps you understand the planet on which we live.

The study of water and the ocean includes the water cycle, water conservation, and the ocean floor. This information helps us use water resources wisely.

Earth science involves the solid outer and liquid inner parts of Earth. By studying rocks, we can learn about Earth's history. This information can be used to find valuable gems, minerals, and other resources.

The study of the atmosphere is part of Earth science. Knowing about weather helps people plan their activities. Predicting storms can save lives and money.

Another aspect of Earth science is the study of space. You may have seen photos taken by satellites and space shuttles. The photo on this page was taken from space. Learning about space and our solar system gives us a better understanding of our own planet, Earth.

Earth scientists are people who study Earth science. Astronomers study the planets, stars, and space. Geologists study rocks and gems. Oceanographers study the seas. June Bacon-Bercey is a meteorologist, a scientist who studies the weather.

During her career, Bacon-Bercey has done many things. She has analyzed and predicted weather for New York, Connecticut, Pennsylvania, and New Jersey. She has been a television weather reporter in Buffalo, New York. She has also worked with pilots to help them with weather forecasts. Bacon-Bercey has worked for the National Oceanic and Atmospheric Administration (NOAA). For NOAA, she teaches people about weather by making films that are shown on television.

Meteorologists study the conditions in the Earth's atmosphere, especially for making weather forecasts.

This unit features articles about many aspects of Earth science.

- Articles about the atmosphere include reading a weather map and an explanation of the greenhouse effect.
- An article about water resources explains why there are water shortages even though three fourths of Earth is water.
- An article about gems explains how these beautiful stones are formed.
- An article about the solar system describes our closest neighbors in space.

Weather

Setting the Stage

Weather moves from one place to another. One day's snow in Kansas City may be the next day's snow in Cincinnati. Weather maps show the weather over large areas. Predictions about the coming weather are based on weather maps.

Past: What you already know

You may already know something about weather or weather maps. Write three things you already know.

1. ______________________

2. ______________________

3. ______________________

Present: What you learn by previewing

Write the headings from the article on pages 101–103 below.

Weather Maps

4. ______________________

5. ______________________

6. ______________________

7. ______________________

What does the map on page 102 show?

8. ______________________

Future: Questions to answer

Write two questions you expect this article to answer.

9. ______________________

10. ______________________

Check your answers on pages 221–222.

Weather Maps

As you read each section, circle the words you don't know. Look up the meanings.

Have you ever noticed that the weather can change overnight? One day is hot, humid, and drizzling, and the next day is clear and dry. It feels as if the air has changed. In fact, the air *has* changed. One large body of air has replaced another.

Air Masses

Large areas of air near Earth's surface take on the same temperature and moisture as the surface. For example, the air over a tropical ocean becomes warm and humid. A large body of air with similar temperature and moisture is called an **air mass**.

There are four types of air masses. The name of an air mass tells you where it came from. The name *continental* refers to a continent. The name *maritime* refers to the sea.

- **Continental polar air masses** are cold and dry. They form over Canada and the northern United States.
- **Continental tropical air masses** are warm and dry. They form over the southwestern United States.
- **Maritime polar air masses** form over the northern Atlantic Ocean and the northern Pacific Ocean. These air masses are cold and moist.
- **Maritime tropical air masses** form over the Caribbean Sea, the middle of the Atlantic Ocean, and the middle of the Pacific Ocean. These air masses are warm and moist.

Air masses do not stay where they form. They may move thousands of miles. Think of a moving air mass as a large, flattened bubble of air. In the United States, air masses are usually pushed from west to east by winds. As the air mass moves, it may keep nearly the same temperature and moisture. You may have noticed that the weather sometimes stays the same for days. That is because the air mass over a location is so large that days go by before it passes.

Fronts

The weather changes when one air mass moves out and another moves in. The leading edge of a moving air mass is called a **front**. A **cold front** is at the front of a cold air mass. A **warm front** is at the front of a warm air mass. When air masses stop moving for a while, the zone between them is called a **stationary front**. The weather can change quickly when a front passes. A front often brings rain or snow to an area.

What Does a Weather Map Show?

Most newspapers print a weather map each day. A **weather map** shows where cold, warm, and stationary fronts are. It also shows temperature and types of **precipitation**, such as rain, snow, and sleet.

Weather maps show areas of high and low pressure. These areas of pressure are important because certain types of weather go with each. Most of the time, a high-pressure area means fair weather and no clouds. A low-pressure area is often cloudy with rain or snow.

High Temperatures and Precipitation for June 28

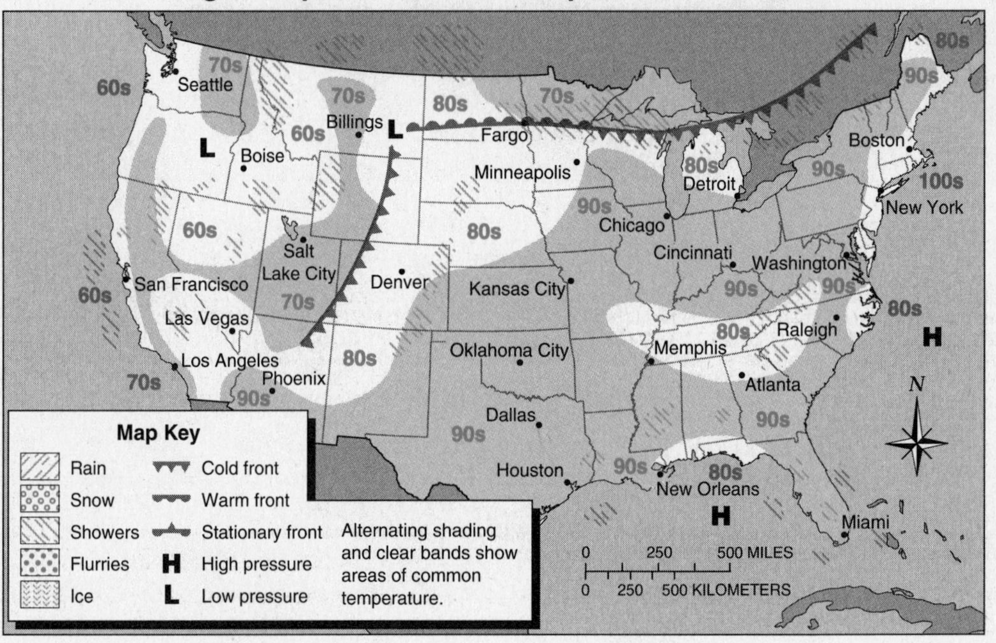

Reading a Map. When you read a weather map, look at the map key to see how information is shown. For example, cold fronts are shown by a line of triangles. The triangles point in the direction in which the front is moving. What is the symbol for a warm front?

Applying Your Skills and Strategies

____

What city is closest to a warm front in the map above?

____

A cold air mass is behind a cold front. What are the temperatures in the cold air mass over the western United States?

____

What are the temperatures in the warm air mass that covers the Midwest and the East?

____

Check your answers on page 222.

Weather Forecasts

Meteorologists are scientists who study the weather. Meteorologists study maps of present weather conditions. Then they decide where the air masses and fronts will be the next day. From this data, they **forecast**, or predict, the next day's weather. Suppose a Dallas meteorologist studied the June 28 map. The forecast might have said more temperatures in the 90s and no rain. Look at the map for June 29 below. Do you think the forecast was correct?

High Temperatures and Precipitation for June 29

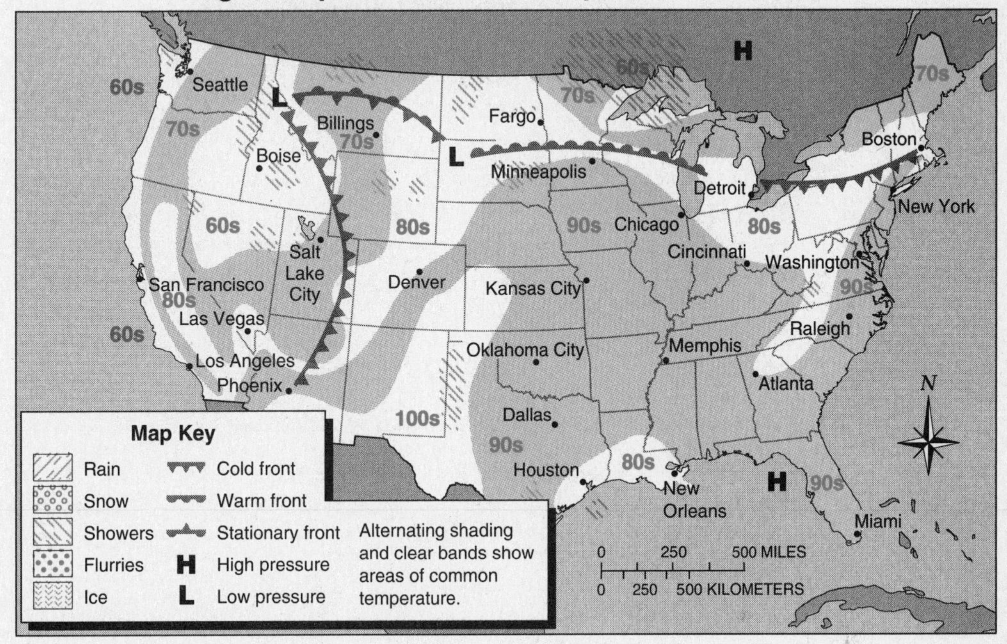

Applying Your Skills and Strategies

Making Predictions. Like a meteorologist, you can use what you know to predict what will happen to the weather. Look at the map for June 28. In the upper right, there is a cold front moving south toward the northeastern United States. Now look at the map for June 29. Circle the cold front.

What is the weather like in New York City on June 29?

____

What kind of weather would you predict for New York City on June 30?

____

People make jokes about the accuracy of weather forecasts. Yet meteorologists are pretty good at predicting tomorrow's weather. However, their long-range predictions are not so accurate. Many factors can affect the weather. So forecasting more than two days in advance involves as much guessing as predicting.

Check your answers on page 222.

Thinking About the Article

Match the characteristics of an air mass with its name. Write the letters of the characteristics in the blank at the left. There are two answers for each item.

	Air Mass	**Characteristic**
______	1. continental polar	a. cold
______	2. continental tropical	b. warm
______	3. maritime polar	c. dry
______	4. maritime tropical	d. moist

Fill in the blank with the word or words that best complete each statement.

5. A large body of air with similar temperature and moisture is called an ________________.

6. The edge of a moving air mass is called a ____________.

7. Scientists who study weather are called ____________________________.

8. A weather ________________ is a prediction of the coming weather.

Write your answers in the space provided.

9. Review the questions you wrote on page 100. Did the article answer your questions? If you said *yes*, write the answers. If your questions were not answered, write three things you learned from this article.

____

____

____

____

____

10. In what direction does weather in the United States generally move?

____

Check your answers on page 222.

Circle the number of the best answer.

11. On a weather map, there is a line of alternating triangles and half circles pointing in opposite directions. What does this indicate?

    (1) cold front

    (2) warm front

    (3) stationary front

    (4) high-pressure area

    (5) precipitation

12. Refer to the map on page 102. In which city are people most likely to be going to the beach?

    (1) Seattle

    (2) San Francisco

    (3) Los Angeles

    (4) New York

    (5) Fargo

13. Refer to the map on page 103. What is the weather forecast likely to be for Salt Lake City?

    (1) occasional showers, high temperature in the 60s

    (2) occasional showers, high temperature in the 90s

    (3) clear, high temperature in the 90s

    (4) clear, high temperature in the 50s

    (5) heavy rain, high temperature in the 80s

Write your answers in the space provided.

14. Refer to the map on page 102. What type of cold air mass is covering the western states? Give reasons for your answer.

____

____

____

15. Describe today's weather in your area. What kind of air mass is in your area now?

____

____

Check your answers on page 222.

The Atmosphere

Setting the Stage

The air that surrounds Earth does more than provide oxygen for us to breathe. It acts like a blanket that helps keep Earth warm. Some scientists think that the air is getting warmer. They think Earth's weather may change.

Past: What you already know

You may already know something about air or how Earth is warmed. Write three things you already know.

1. ______________________________

2. ______________________________

3. ______________________________

Present: What you learn by previewing

Write the headings from the article on pages 107–109 below.

The Greenhouse Effect

4. ______________________________

5. ______________________________

6. ______________________________

What does the diagram on page 107 show?

7. ______________________________

Future: Questions to answer

Write three questions you expect this article to answer.

8. ______________________________

9. ______________________________

10. ______________________________

Check your answers on page 222.

The Greenhouse Effect

As you read each section, circle the words you don't know. Look up the meanings.

The 1980s were unusually warm. The four hottest years on record occurred in that decade. Is the recent heat wave part of a rise in Earth's temperature? Or were the warm years just a matter of chance? Whatever caused this trend, the heat has focused people's attention on global, or worldwide, warming.

How Earth Is Warmed

The air around us, called the **atmosphere**, plays a large role in the warming of Earth. When the gases in the atmosphere absorb energy, they become warmer. But from where does the energy come? It comes from two places, the sun and Earth.

Energy from the sun is called **radiant energy**. When you are outside on a bright day, the radiant energy of sunlight warms you. The atmosphere absorbs about 20 percent of the sun's radiant energy and reflects about 30 percent back into space. The remaining 50 percent of the sun's radiant energy is absorbed by Earth.

Earth radiates energy back into the atmosphere as **infrared radiation**. We feel infrared radiation as heat. You can feel heat rising from the pavement or from sand on a beach. These are examples of infrared radiation.

The infrared radiation reflected from Earth's surface does not escape into space. Instead, carbon dioxide, water vapor, and other gases in the atmosphere trap the heat. This is called the **greenhouse effect** because it is similar to what happens in a greenhouse.

The Greenhouse Effect

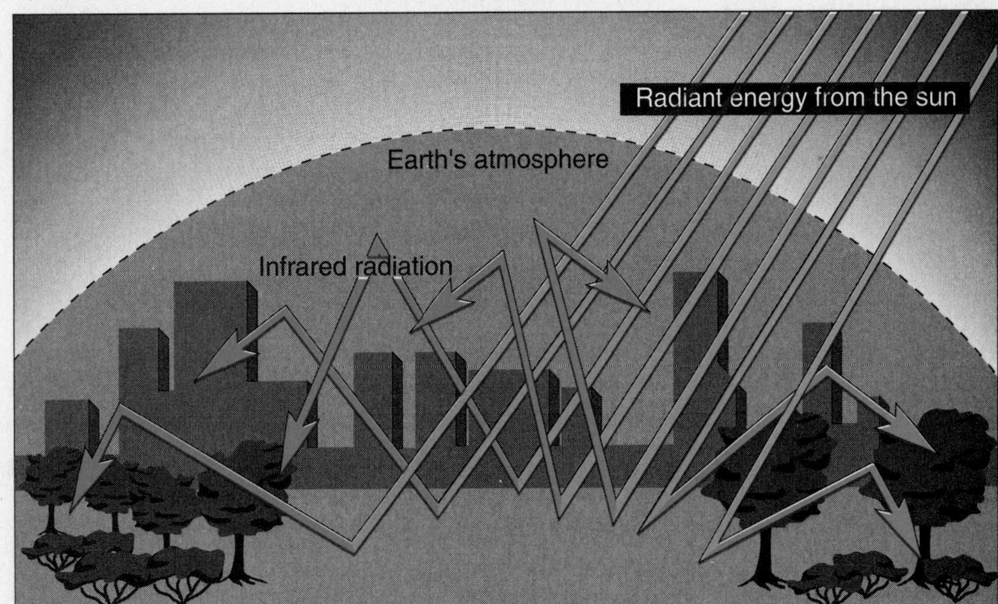

Are People's Activities Increasing the Greenhouse Effect?

The greenhouse effect is a normal effect of the atmosphere. Without it, Earth would be much colder. But scientists say that the blanket of gases that absorbs infrared radiation is getting thicker. This may cause Earth to become warmer. Such a change would be the result, or effect, of human activity.

Applying Your Skills and Strategies

Understanding Compound Words. Science books are full of long words. But most of these words are made of smaller parts. Sometimes you can figure out what a long word means if you know what each part means. Compound words are made by combining two smaller words. Some familiar compound words are *baseball*, *sunshine*, and *supermarket*.

There are several compound words in the article you are reading. A *greenhouse* is a building (house) for growing plants (green). A *landfill* is a place to bury trash (fill) in the ground (land). What is sunlight?

People are changing the greenhouse effect. In the last hundred years, people have burned more and more fossil fuels. Fossil fuels include coal, oil, gasoline, and wood. When they burn, they increase the amount of carbon dioxide in the air. Also, people have destroyed many forests throughout the world. Plants absorb carbon dioxide during photosynthesis. Fewer plants means that more carbon dioxide remains in the atmosphere. More carbon dioxide in the atmosphere means that more heat is absorbed.

Carbon dioxide is the main "greenhouse gas," but there are several others. Ozone, chlorofluorocarbons (CFCs), methane, and nitrogen oxide all absorb infrared radiation. These greenhouse gases have increased as a result of pollution from factories and farms.

Sources of greenhouse gases

Check your answer on page 222.

Scientists estimate that Earth's temperature has increased between 0.9 and 2.7°F in the last 150 years. Some think it could go up another 8°F by the year 2030. Global warming could melt the polar ice caps and flood coastal areas. It could reduce the amount of land suitable for farming.

Applying Your Skills and Strategies

Understanding the Relationships Among Ideas. When you read, you are thinking all the time. Your mind is busy linking ideas to one another and to things you already know.

Reread the section of the article under the heading *Are People's Activities Increasing the Greenhouse Effect?* The main ideas are: (1) People are burning more fossil fuels. (2) People are destroying forests. (3) Pollution is increasing. (4) Carbon dioxide and other greenhouse gases are increasing. (5) Global temperature is increasing.

The first three ideas are related. They are all things people do that cause an increase in greenhouse gases. Ideas 4 and 5 are related. What do they have in common?

____

What is the relationship of ideas 1, 2, and 3 to ideas 4 and 5?

____

What Can Be Done About the Greenhouse Effect?

The problem of global warming may seem so large that no one can do anything about it. But there are many things an individual can do to help. Almost anything a person does that saves energy means that less fossil fuels are burned. If less fossil fuels are burned, there is less pollution.

Energy can be saved at home by adding insulation and turning down the thermostat. Keeping the furnace in good repair will cut down the amount of carbon dioxide it produces. Conserving electricity, which is usually produced by burning fossil fuels, will help.

Energy can be saved by driving a car that gets many miles per gallon of gasoline. Driving only when necessary will help, too. Make sure the car's air conditioner is in good working order. Air conditioners can leak CFCs into the air.

Recycling also saves energy. When recycled materials are used in manufacturing, less energy is used. Also, recycling can help reduce the amount of garbage in landfills and incinerators. Landfills produce methane gas when garbage breaks down. Garbage burned in incinerators produces carbon dioxide.

In addition, each person can plant a tree. A single tree may not do much, but it will help. Some cities, such as Los Angeles, are even sponsoring tree plantings.

Check your answers on page 222.

Thinking About the Article

Fill in the blank with the word or words that best complete each statement.

1. The air that surrounds Earth is called the ______________________.
2. Energy from the sun is called ______________ energy.
3. The heat you feel radiating up from hot pavement is called

   ______________________________________.
4. Gases in the atmosphere absorb energy radiated from Earth. This is

   known as the ______________________________.

Write your answers in the space provided.

5. Review the questions you wrote on page 106. Did the article answer your questions? If you said *yes*, write the answers. If your questions were not answered, write three things you learned from this article.

____

____

____

____

____

6. What kind of energy warms you when you sit out in the sun?

____

____

7. Why is the greenhouse effect like a blanket?

____

____

8. List the major greenhouse gases.

____

____

Check your answers on pages 222–223.

Circle the number of the best answer.

9. The amount of carbon dioxide in the atmosphere has been increased by
   (1) burning coal.
   (2) burning wood.
   (3) burning oil.
   (4) destroying forests.
   (5) all of the above.

10. Which of the following reduces the level of carbon dioxide in the atmosphere?
    (1) photosynthesis
    (2) air pollution
    (3) greenhouse effect
    (4) burning fossil fuels
    (5) cutting down trees

11. Why would sea levels rise if the temperature on Earth becomes several degrees warmer?
    (1) The ocean would expand because of the extra heat.
    (2) Some land would no longer be suitable for farming.
    (3) The polar ice caps would melt.
    (4) There would be more waves.
    (5) The pull of Earth's gravity would decrease.

Write your answers in the space provided.

12. Think about the weather in your area over the last few years. Has the weather been unusual in any way? Explain your answer.

____

____

13. What specific things can you do to help reduce the greenhouse effect?

____

____

____

Check your answers on page 223.

Resources

Setting the Stage

Earth has plenty of water. However, fresh water is not always available where it is needed. Some places must get water from great distances. In many areas, there is not enough water for everyone.

Past: What you already know

You may already know something about water or saving water. Write three things you already know.

1. ______________________________

2. ______________________________

3. ______________________________

Present: What you learn by previewing

Write the headings from the article on pages 113–115 below.

Fresh Water

4. ______________________________

5. ______________________________

6. ______________________________

What does the circle graph on page 113 show?

7. ______________________________

Future: Questions to answer

Write three questions you expect this article to answer.

8. ______________________________

9. ______________________________

10. ______________________________

 Check your answers on page 223.

Fresh Water

As you read each section, circle the words you don't know. Look up the meanings.

In Santa Barbara, California, water use has been cut by half. Lawn watering has been banned. Some people are painting their dry grass green. In San Francisco, homes and businesses can use only 75 percent of their normal amount of water. Those who use more are fined. Los Angeles households must cut back water use by 10 percent.

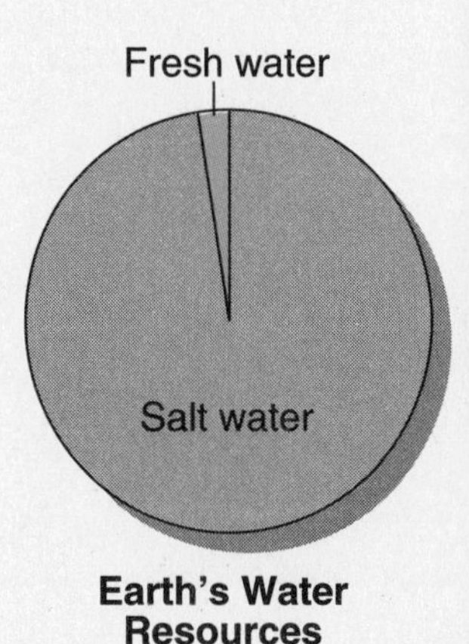

Earth's Water Resources

Dried up Nicosia Reservoir in Marin County, California

California and other western states have been suffering through a drought. Cities do not have enough water. But in the San Joaquin, Sacramento, and Imperial valleys, farms are green. Alfalfa, cotton, and rice are growing very well. These crops, as well as livestock, are grown on thousands of acres of **irrigated**, or watered, land. Farmers use 85 percent of California's water to grow crops.

California, like many areas of the West, is mostly desert. Normal rainfall in Los Angeles is 14 inches per year—about the same as Tripoli, Libya. San Francisco gets less precipitation than Tel Aviv, Israel. Without water from other places, California could not support its farm industry or its 23 million people.

Water as a Resource

Astronauts often call Earth the blue planet because it is covered by water. Earth has plenty of water, but most of it is in the oceans. Ocean water is salty, and you cannot drink it. Less than 3 percent of Earth's water is fresh water. Most of this fresh water is frozen in ice at the poles. About 1 percent of all Earth's water is available as a resource. A **resource** is a substance that is needed. Water is a resource that is found underground in **groundwater**. It is also found in rivers, lakes, and the atmosphere.

One percent may seem like a very small amount of fresh water, but it is not. For one thing, the supply of fresh water is constantly being replaced. Water is a **renewable resource**, so it's one that doesn't get used up. In the United States, enough precipitation falls to provide each person with 5,470 gallons of water per day. Do you use that much water? Of course not. In fact, each person uses about 1,650 gallons each day. That includes water used in homes, schools, farms, and businesses.

Applying Your Skills and Strategies

Drawing Conclusions. A *conclusion* is an idea that follows logically from the information you have. Conclusions must be supported by *facts*. You have read that the United States, as a whole, receives more than enough rain for the needs of all its people. Yet in the western states, there is not enough water to go around. You might conclude that the West is having problems because it receives less rainfall than the rest of the United States. Write two other conclusions based on the facts you have learned so far.

____

____

Water Supplies

The United States may have more than enough water for everyone. However, one problem is that the water is not always where it is needed. For example, California uses water from miles away. Most precipitation in the West falls in mountains, such as the Cascades and the Sierra Nevada. It flows as surface water in rivers, such as the Sacramento and Colorado. **Reservoirs**, lakes created by dams, store water. Pipes called **aqueducts** carry the water to southern California.

California's Water Supply System

Check your answers on page 223.

Another problem is that supplies of surface water depend on precipitation. If not enough rain or snow falls for a long period of time, a drought occurs in that area. Groundwater, rivers, lakes, and reservoirs become low. During a drought, water use may have to be restricted.

Groundwater supplies are threatened by overuse and pollution. When water is taken from a well, the groundwater level goes down. In West Texas, water for irrigation has been pumped for almost one hundred years. In that time, the level of the groundwater has dropped almost one hundred feet. In other areas, polluted wells have been shut down.

Finally, as the population of an area grows, the need for water increases. Southern California's population has been growing since the early 1900s. More and more people are moving to the cities. Some people in California feel that the state has enough water. But they think that farms are getting more than their fair share. The farmers and their supporters argue that California supplies most of the country with food products. These people feel that irrigated open land adds to the quality of life for everyone in the state. While California tries to solve its long-term water problems, its citizens must conserve water.

Applying Your Skills and Strategies

Distinguishing Fact from Opinion. Remember that *facts* can be proved true. *Opinions* are beliefs that may or may not be true. In the previous paragraph, you read different opinions about California's water problems. Circle the sentences in the paragraph that express opinions.

What do you think about California's water problems? Do you think city people should conserve water so that farms can be irrigated? Write your opinion below.

___

___

Conserving Water

Many areas, not just California, have problems with their water supply. That is why it's important to get into the habit of conserving water. Here are some things you can do.

- Take shorter showers. It is better to take a bath rather than a shower. A full tub uses half the water of a ten-minute shower.
- Fix leaks and running toilets. A dripping faucet can waste three hundred to six hundred gallons of water per month.
- Run the washing machine and dishwasher only with full loads.
- If you're doing dishes in the sink, don't run the water. Use one full basin to wash and another to rinse.
- Turn off the water when you're brushing your teeth or shaving. A bathroom faucet uses up to five gallons of water per minute.

Check your answers on page 223.

Thinking About the Article

Fill in the blank with the word or words that best complete each statement.

1. In areas with low levels of rainfall, farms must be

   ________________ in order to grow crops.

2. A ________________ is a substance that is needed.

3. Water that is found underground is called ________________.

4. Water is a ______________________________, so it doesn't get used up.

5. A ________________ is a lake created by a dam.

6. A pipe that carries water long distances is an ________________.

Write your answers in the space provided.

7. Review the questions you wrote on page 112. Did the article answer your questions? If you said *yes*, write the answers. If your questions were not answered, write three things you learned from this article.

8. How is most of the water in California used?

9. Why have cities in California started conserving water?

10. Where is fresh water found?

Check your answers on page 223.

Circle the number of the best answer.

11. According to the article, the amount of rainfall in Los Angeles is about the same as the amount of rainfall in
    (1) New York City.
    (2) Miami.
    (3) Tripoli.
    (4) London.
    (5) Bombay.

12. Why are there water shortages in parts of the United States?
    (1) Not enough rain falls in the United States each year.
    (2) All of the water in the United States is polluted.
    (3) All of the water in the United States is saltwater.
    (4) Some areas get more precipitation than they need, while others do not get enough.
    (5) The amount of water used in homes is increasing.

13. According to the article, water is a renewable resource. Which of the following is also a renewable resource?
    (1) sunlight
    (2) oil
    (3) coal
    (4) natural gas
    (5) copper

Write your answers in the space provided.

14. Where does the water in your community come from?

____

15. Do you think that people in your community would volunteer to conserve water during a drought?

____

16. Describe ways in which you or someone you know has conserved water.

____

____

Check your answers on page 223.

Minerals

Setting the Stage

Mineral crystals are among the most beautiful forms in nature. They have a regular structure and shape. The conditions under which crystals form affect their size. Minerals, including gemstones, are used for many things.

Past: What you already know

You may already know something about crystals or minerals. Write three things you already know.

1. ______________________________

2. ______________________________

3. ______________________________

Present: What you learn by previewing

Write the headings from the article on pages 119–121 below.

Beautiful Crystals

4. ______________________________

5. ______________________________

6. ______________________________

What does the diagram on page 120 show?

7. ______________________________

Future: Questions to answer

Write three questions you expect this article to answer.

8. ______________________________

9. ______________________________

10. ______________________________

 Check your answers on page 224.

Beautiful Crystals

As you read each section, circle the words you don't know. Look up the meanings.

Crystals have appealed to people's imaginations for thousands of years. Some crystals, such as emeralds, were thought to cure sickness. Others, such as rubies, were believed to bring luck and protection from a violent death. The ancient Greeks thought that crystals were made of ice frozen so hard it would never thaw. Some crystals are so famous they have names. The Koh-i-noor, Star of India, and Hope diamonds are a few famous crystals.

What Is a Crystal?

Before we find out what a crystal is, we have to understand what a mineral is. A **mineral** is a solid substance that has a specific composition and structure. Gold, copper, lead, diamonds, and salt are examples of minerals.

Minerals, like all other matter, are formed of tiny particles called **atoms**. Each different kind of atom is called an element. An **element** is a substance that cannot be changed by chemical means into a simpler substance. About ninety elements occur naturally on Earth. Iron and lead are examples of elements. An element also can combine with another element to form a new substance called a **compound**. Minerals are either elements or compounds. So far, over two thousand kinds of minerals have been identified.

The head of the State Sceptre contains part of the Cullinan Diamond known as the "Second Star of Africa."

The Hope Diamond

The atoms of most minerals, whether they are elements or compounds, are packed together in a regular pattern, not a jumble. The result is a **crystal**, a solid with a regular shape and flat sides. The shape of a crystal depends on the way its atoms are arranged. In some minerals, the crystals are so small they can be seen only with a microscope. Other minerals, such as quartz, have large crystals.

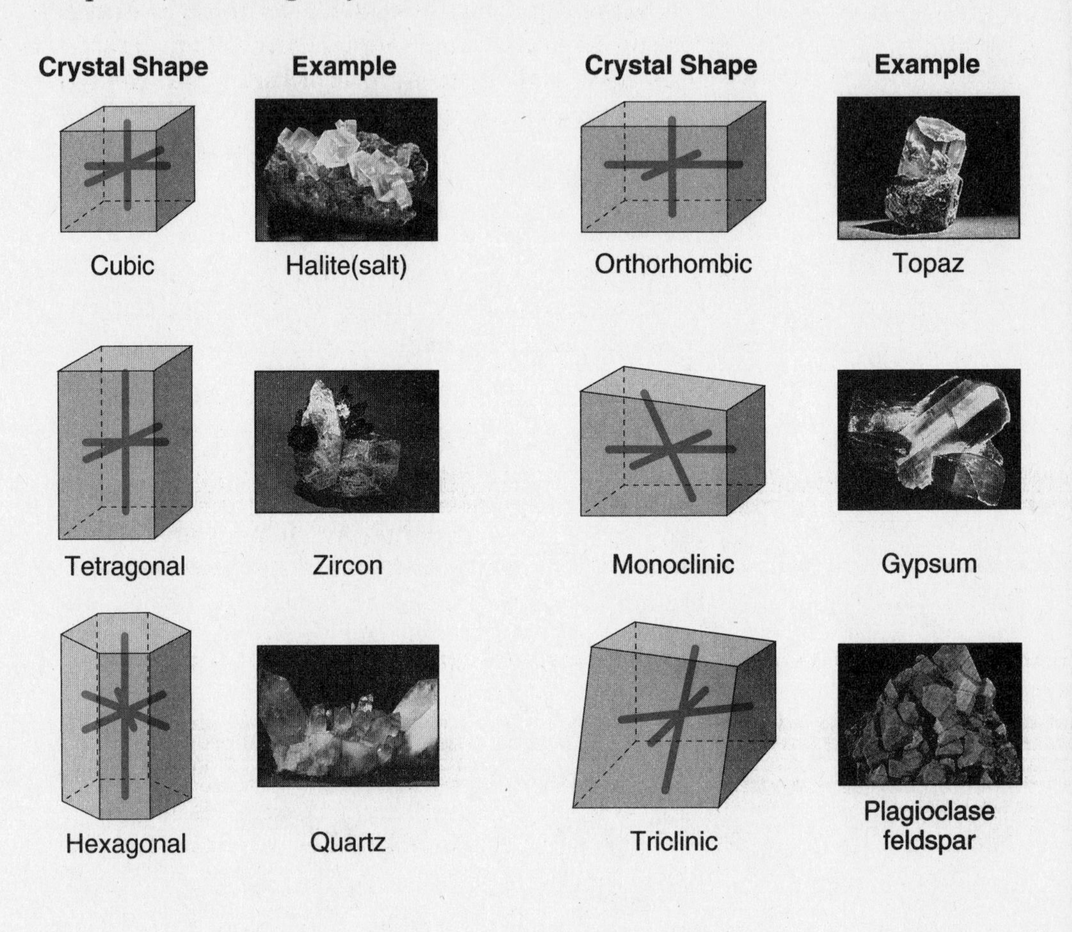

Finding Details. Facts, observations, and measurements are important details in science. Often details are found in charts and diagrams. That is why it's important to look carefully at charts and diagrams when you read.

Applying Your Skills and Strategies

The first paragraph on this page tells that crystals have structure and shape. To learn about the different structures and shapes, you must look at the diagram above. Circle the names of the six basic crystal shapes shown on the diagram. Which crystal shape looks like a cube?

____

Which crystal shape looks like a tube with six sides?

____

Give an example of a mineral with monoclinic structure.

____

Check your answers on page 224.

How Crystals Are Formed

Crystals can form in different ways. When hot liquid rock or metal cools and becomes solid, crystals are formed. As the atoms cool, they become locked into a pattern that grows. If the cooling is fast, the crystals are small. If the cooling is slow, the crystals are large.

Some liquids have minerals that are dissolved in them. When a mineral solution evaporates, crystals form. For example, when seawater evaporates, salt crystals form. People make rock candy by dissolving sugar in water, then letting the water evaporate.

Applying Your Skills and Strategies

Diagramming Ideas. Thinking about how one idea relates to another helps you understand what you read. Sometimes drawing a diagram to show the important ideas makes the relationships among them clearer. For example, you can diagram the important ideas from the previous two paragraphs as shown below.

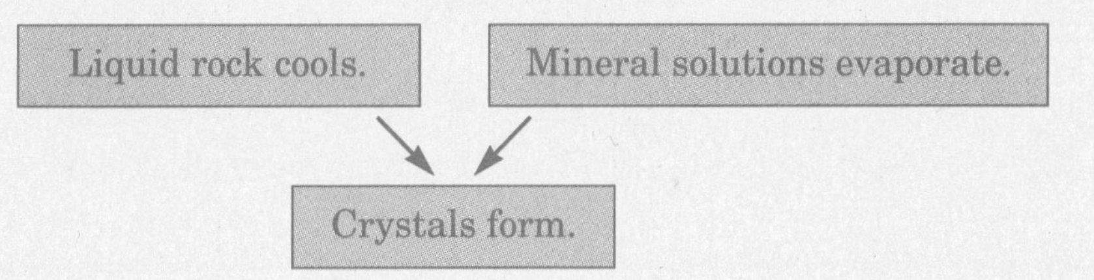

Reread the material on page 119 about elements, compounds, and minerals. In the following diagram, finish filling in the boxes to show the relationships among elements, compounds, and minerals.

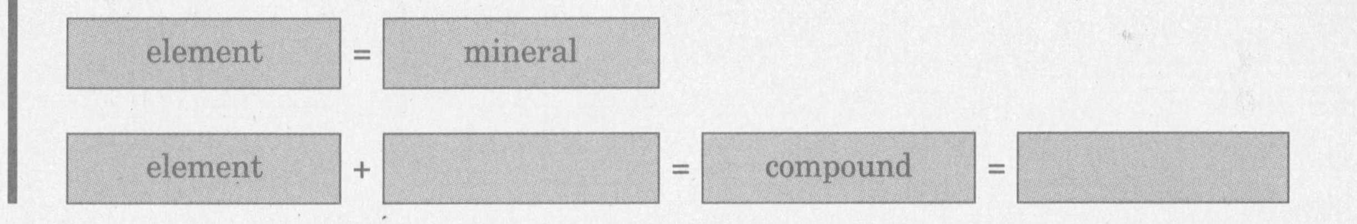

All crystals are different because they form in different ways. They may grow together in odd shapes. If there is not enough space for crystals to form, their shapes may be distorted.

The Importance of Minerals

Rocks containing minerals that can be mined for profit are called **ores**. Ores that contain metals are called **metallic ores**. Silver, gold, iron, copper, and aluminum come from metallic ores. **Nonmetallic ores** contain minerals that are not metals. Sulfur is an example of a mineral that comes from a nonmetallic ore.

Some minerals are valued for their beauty, durability, or rarity. **Gems** are colorful minerals that often are cut, polished, and made into jewelry. Diamonds, sapphires, amethysts, opals, and garnets are a few gems.

Check your answer on page 224.

Thinking About the Article

Fill in the blank with the word or words that best complete each statement.

1. A ____________ is a solid substance that has a specific composition and structure.

2. All matter is formed of tiny particles called ____________.

3. A ____________ is a substance formed when elements combine with one another.

4. A ____________ is a solid with a regular shape and flat sides.

5. ____________ are rocks containing minerals that can be mined for profit.

6. ____________ are rocks that contain metals.

7. ____________ are colorful minerals valued for their beauty, durability, or rarity.

Write your answers in the space provided.

8. Review the questions you wrote on page 118. Did the article answer your questions? If you said *yes*, write the answers. If your questions were not answered, write three things you learned from this article.

____

____

____

____

____

9. Why do crystals have a regular shape?

____

____

____

Check your answers on page 224.

Circle the number of the best answer.

10. Refer to the diagram on page 120. Which crystal shape does quartz have?

    (1) cubic

    (2) tetragonal

    (3) hexagonal

    (4) orthorhombic

    (5) monoclinic

11. According to the diagram on page 120, which of the following minerals has the cubic crystal shape?

    (1) halite

    (2) zircon

    (3) topaz

    (4) gypsum

    (5) plagioclase feldspar

12. How are crystals formed?

    (1) when liquid containing dissolved minerals evaporates

    (2) when liquid rock or metal cools and becomes solid

    (3) when atoms become arranged in a jumble

    (4) options 1 and 2 only

    (5) options 2 and 3 only

Write your answers in the space provided.

13. Draw a diagram that shows the relationships among ore, metallic ore, and nonmetallic ore.

14. Minerals may be food, metals, or gems. Describe the minerals you own or some that you would like to own.

___

___

___

Check your answers on page 224.

The Solar System

Setting the Stage

Two unmanned space probes, named *Voyager 1* and *Voyager 2*, did a grand tour of four planets in 12 years. By sending photos and data to Earth, these space probes gave us a closer look at Jupiter, Saturn, Uranus, and Neptune.

Past: What you already know

You may already know something about space exploration or the planets. Write three things you already know.

1. ______________________________

2. ______________________________

3. ______________________________

Present: What you learn by previewing

Write the headings from the article on pages 125–127 below.

The Outer Planets

4. ______________________________

5. ______________________________

6. ______________________________

7. ______________________________

What does the diagram on page 125 show?

8. ______________________________

Future: Questions to answer

Write two questions you expect this article to answer.

9. ______________________________

10. ______________________________

Check your answers on page 224.

The Outer Planets

As you read each section, circle the words you don't know. Look up the meanings.

The first spacecraft was launched in the 1950s. Since then, scientists have been learning new things about the solar system. The **solar system** is made up of the sun and the objects that revolve around the sun. These objects include the planets and their moons. Mercury and Venus are called the **inner planets** because they are between Earth and the sun. Mars, Jupiter, Saturn, Uranus, Neptune, and Pluto are called the **outer planets**. They are farther from the sun than Earth.

The Mission

Most space probes have been designed to visit only one planet. But scientists got a bonus with *Voyager 1* and *Voyager 2*. When these probes were launched, four of the outer planets were on the same side of the sun. These planets would not be in this position again for 175 years. So with only enough fuel to get to Jupiter, the *Voyager* space probes flew on to Saturn, Uranus, and Neptune. From Jupiter on, the space probes used the gravity of one planet to speed on to the next. Despite problems with radio reception, cameras, and computers, scientists on Earth guided *Voyager 1* and *Voyager 2* through an almost perfect grand tour.

The two space probes were launched in 1977. For the visits to Jupiter and Saturn, *Voyager 1* was the main spacecraft. *Voyager 2* was the backup in case *Voyager 1* failed. After *Voyager 1* traveled to Jupiter and Saturn, it headed out of the solar system. *Voyager 2* continued on to visit Uranus and Neptune.

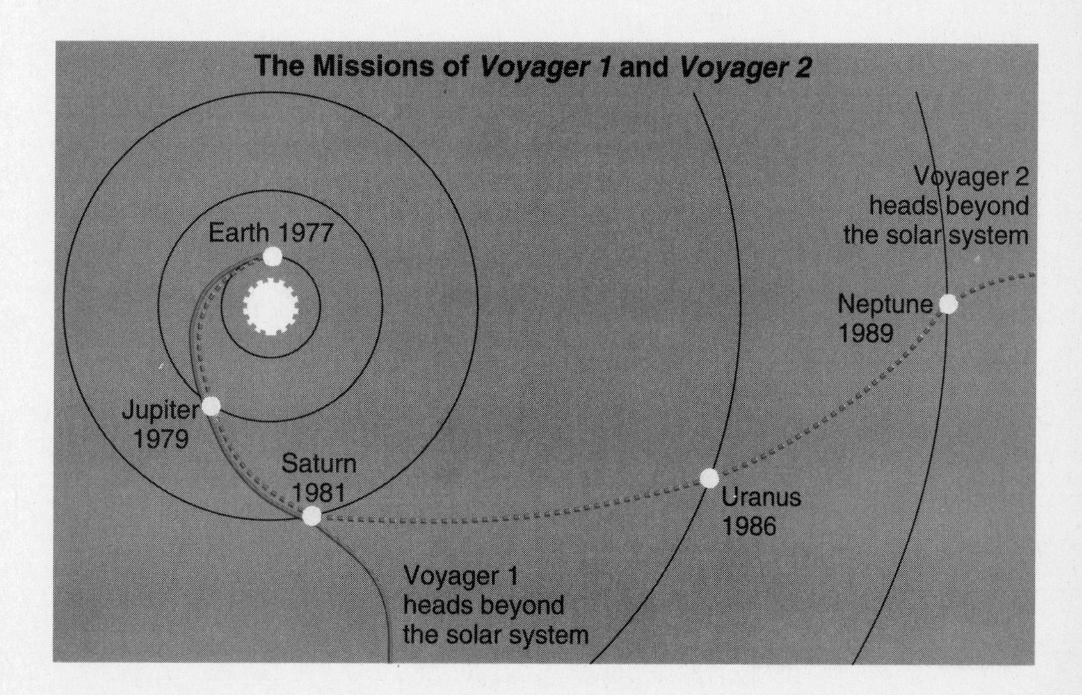

Between 1979 and 1989, *Voyager 1* and *Voyager 2* sent back spectacular photos. They also sent back data about Jupiter, Saturn, Uranus, Neptune, and their moons. Some basic data about these four planets and Earth are shown here.

Earth and Four Outer Planets

	Earth	Jupiter	Saturn	Uranus	Neptune
Diameter (miles)	8,000	89,000	75,000	32,000	30,000
Mass (compared to the mass of Earth)	1	318	95	15	17
Distance from the sun (millions of miles)	93.5	486.4	892	1,790	2,810
Time of one revolution around sun (years)	1	12	29	84	165
Time of one rotation (length of day in hours)	24	10	11	16	18
Number of known moons	1	16	17	14	8

Applying Your Skills and Strategies

Reading a Table. One way to present a set of facts is to organize them in a table or chart. When you read a table, start with the title to get the main idea. Circle the title of the table above. Find the headings over each vertical column. Draw a box around each heading. The table includes several characteristics of planets. Underline each characteristic. Then use the table to answer the following questions.

How many moons does Jupiter have?

___

How long does it take Uranus to revolve around the sun?

___

Jupiter photographed by *Voyager 1* when it was 22 million miles away from the planet

Jupiter and Saturn

Voyager 1 passed Jupiter in 1979. The photos that were sent back showed that Jupiter has rings. Jupiter also has a swirling, stormy atmosphere made mostly of hydrogen and helium. The Great Red Spot of Jupiter is a storm several times larger than Earth.

Jupiter's four largest moons were first seen by Galileo in 1610. *Voyager 1* provided a closer look. Ganymede, the largest moon in the solar system, looks similar to our moon. Io has active volcanoes, and its surface looks like pizza. Europa is smooth, with lines on its surface. Callisto looks old and has deep craters.

Check your answers on page 225.

Saturn's rings photographed by *Voyager 2* when it was 27 million miles from Saturn

Next, *Voyager 1* went to Saturn, the second largest planet. Like Jupiter, Saturn is made mostly of hydrogen and helium. Saturn rotates quickly, causing bands of clouds to form in its atmosphere. Wind speeds of three hundred miles per hour were measured.

The biggest surprise for scientists was Saturn's rings. Instead of there being a few rings, it was discovered that the planet has thousands of rings. Saturn's rings are made of particles ranging from specks of dust to large rocks. Some rings have "spokes" that appear and disappear. This new information has raised many questions about Saturn's rings.

Uranus and Neptune

It took almost five years for *Voyager 2* to travel the distance between Saturn and Uranus. Uranus is made mostly of hydrogen and helium. The temperature in the atmosphere is about −330°F. At Uranus, *Voyager 2* found 10 new moons, bringing the total number of known moons to 14. In addition, *Voyager 2* showed that Uranus's rings are made up mostly of large rocks with only small amounts of dust.

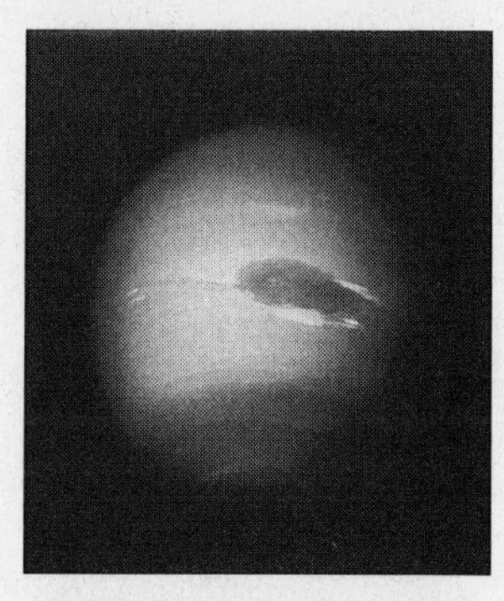

Neptune photographed from *Voyager 2*

The last planet *Voyager 2* flew by was Neptune in 1989. Neptune, made mostly of hydrogen, also has a stormy atmosphere. One feature seen by *Voyager 2* is the Great Dark Spot. This spot is a storm almost as big as Earth. Data sent back to Earth suggests that Neptune's winds, moving over 1,200 miles per hour, might be the fastest in the solar system.

Voyager 2 also discovered six new moons orbiting Neptune, bringing the total number of known moons to eight. The largest moon, Triton, has a diameter of 1,700 miles. The smallest moon has a diameter of 34 miles.

Applying Your Skills and Strategies

Drawing Conclusions. *Conclusions* are ideas that are based on *facts*. They follow logically from the facts. This article describes four of the outer planets and some of their moons. It doesn't come to many conclusions. Based on the information in the article, you can write conclusions about the planets.

What is the relationship between a planet's distance from the sun and the time it takes to revolve around the sun?

____

____

The Mission Continues

With their main mission accomplished, *Voyager 1* and *Voyager 2* continue to speed out of the solar system. Although their cameras no longer take pictures, other instruments send information back to Earth. It is estimated that their power sources will last until the year 2015. After that, they will stop sending data. If aliens come across one of the *Voyager* space probes, they will find a recording aboard. It has greetings from Earth in sixty languages.

Check your answer on page 225.

Thinking About the Article

Match each planet with its description. Write the letter of the planet in the blank at the left.

	Description	Planet
______	1. Largest planet in the solar system	a. Earth
______	2. Planet with the most rings	b. Saturn
______	3. Has longest time of revolution	c. Jupiter
______	4. Rotates in 24 hours	d. Uranus
______	5. Has 14 known moons	e. Neptune

Write your answers in the space provided.

6. Review the questions you wrote on page 124. Did the article answer your questions? If you said *yes*, write the answers. If your questions were not answered, write three things you learned from this article.

____

____

____

____

____

7. What makes up the solar system?

____

____

8. What was the mission of the *Voyager* space probes?

____

____

____

Check your answers on page 225.

Circle the number of the best answer.

9. According to the table on page 126, which planet is about twice as far from the sun as Saturn?
   (1) Earth
   (2) Jupiter
   (3) Uranus
   (4) Neptune
   (5) none of the above

10. According to the table on page 126, which planet has the most known moons?
   (1) Earth
   (2) Jupiter
   (3) Saturn
   (4) Uranus
   (5) Neptune

11. Scientists controlled the systems on *Voyager 1* and *Voyager 2* from Earth. Which of the following is most similar to the way the space probes were controlled?
   (1) setting a timer that starts an appliance
   (2) turning a switch on and off
   (3) mounting a camera on a tripod
   (4) operating a VCR or TV with a remote control
   (5) listening to music through headphones

Write your answers in the space provided.

12. Why is it not practical to send astronauts on missions to the outer planets at this time?

___

___

13. Which of the planets visited by *Voyager 1* or *Voyager 2* interests you the most? Explain your answer.

___

___

Check your answers on page 225.

Unit 2 Review: Earth Science

High Temperatures and Precipitation for June 29

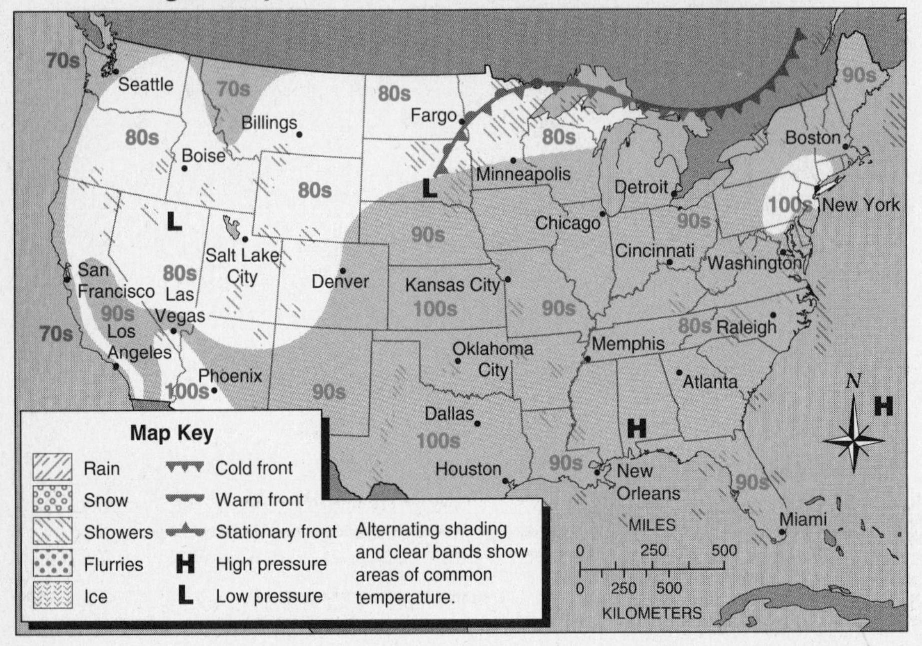

Weather Maps

Most daily newspapers print a weather map. The heavy lines show fronts. The terms *warm front* and *cold front* describe the kind of temperatures that are behind the front. The symbols point in the direction in which the front is moving. A stationary front forms when a warm front and a cold front meet, and neither one moves.

Fill in the blank with the word or words that best complete each statement.

1. According to the key, a series of small circles on the map indicates that an area will have __________.

2. The season in which this map was prepared was __________.

Circle the number of the best answer.

3. What is the weather like along the stationary front?
   (1) hot, with sunshine
   (2) hot, with showers
   (3) cold, with sunshine
   (4) cold, with snow flurries
   (5) moderate temperatures, with high winds

Go on to the next page.

The Ozone Layer

You may have noticed that many products that used to be sold in aerosol spray cans are now sold in pump sprays. The reason for this is that many of the gases used in aerosols harm the environment. These gases are called chlorofluorocarbons (CFCs). CFCs escape into the air when products containing them are used. Once in the atmosphere, CFCs can damage the ozone layer.

Ozone is a form of oxygen. A layer of ozone is found ten to fifty miles above Earth's surface. The ozone layer absorbs ultraviolet rays from the sun. **Ultraviolet rays** cause sunburn and skin cancer. The ozone layer protects plants and animals from the effects of these harmful ultraviolet rays.

The ozone layer has become thinner in the last twenty years. There is now a hole in the ozone layer. It is over Antarctica and seems to be getting larger each year. Many people are concerned about the damage that has been done to the ozone layer. Without a protective layer of ozone, ultraviolet rays may cause increases in the cases of skin cancer throughout the world. Many environmental groups are working to get CFCs banned.

Fill in the blank with the word or words that best complete each statement.

4. Ozone is a form of ____________.

5. ______________________ rays are dangerous rays from the sun.

Circle the number of the best answer.

6. Which of the following statements does not describe the ozone layer?
   (1) It has a hole in it.
   (2) It is ten to fifty miles above Earth's surface.
   (3) It is getting thinner.
   (4) It absorbs ultraviolet rays.
   (5) It is made up of CFCs.

7. If the use of CFCs were to increase, more people would probably die of
   (1) oxygen poisoning.
   (2) CFC poisoning.
   (3) skin cancer.
   (4) sunburn.
   (5) all of the above.

Go on to the next page.

The Oceans

Oceans cover nearly three-fourths of Earth's surface. The water is shallow where the oceans meet the continents. The bottom of the ocean has a gentle slope in this area. It is called the **continental shelf**, because the ocean bottom is almost flat. Since the water is shallow, sunlight can reach the bottom. There are many plants here. There are also many animals. We get shellfish, such as clams and lobsters, from the continental shelf. This is also the richest part of the sea for fishing.

The continental shelf extends from just a few miles to a thousand miles from the shore. Then the bottom slopes more steeply. This is the **continental slope**. The slope levels out to form the **ocean basin**, the bottom of the sea. The resources from these regions come mostly from the upper layers of the water in the open ocean. Large fish, such as tuna, are caught here.

The ocean gives us more than food resources. The rock layers of the continental shelf are sources of oil and natural gas. On the ocean basin are lumps of minerals, called nodules. They contain mostly manganese. They also contain some iron, copper, and nickel.

Fill in the blank with the word or words that best complete the statement.

8. The part of the ocean where the water meets the land is the

    ______________________________.

Circle the number of the best answer.

9. The best food resources in the ocean are in

    (1) nodules on the ocean basin.

    (2) nodules on the continental shelf.

    (3) waters of the open ocean.

    (4) waters of the continental-shelf region.

    (5) waters of the continental-slope region.

10. All of the following are resources from the ocean except

    (1) oil.

    (2) manganese.

    (3) iron.

    (4) natural gas.

    (5) coal.

The Sun

The sun is the nearest star to Earth. The sun appears very large and bright. However, it is not large or bright when compared to other stars. The sun appears large and bright because it is so much closer than any other star. The next nearest star is more than 250,000 times farther away.

Huge amounts of energy are given off by the sun. The energy comes from nuclear reactions in the sun. These reactions take place in the **core**, which is the center of the sun. The temperature there is believed to be about 27 million°F. A large portion of the heat is changed into light on the surface of the sun. As a result, the temperature of the surface is only about 10,000°F.

Even at this temperature, the sun is very hot. Earth is about 93 million miles from the sun. This is just the right distance for people to live on Earth. If Earth were closer to the sun, it would be too hot to support life. If Earth were much farther from the sun, it would be too cold.

Energy, in the form of light from the sun, is responsible for life on Earth. Plants use the sun's energy to make food. Through food chains, all animals are fed. The sun's energy warms Earth, which causes winds. Winds carry weather systems from place to place.

Fill in the blank with the word or words that best complete each statement.

11. The nearest star to Earth is the __________.
12. The hottest part of the sun is the __________.
13. Plants use the sun's energy to make ________.
14. All animals are fed through ______________________.

Circle the number of the best answer.

15. Many people have solar-powered calculators. What kind of energy from the sun do these devices use?

    (1) heat

    (2) light

    (3) nuclear

    (4) wind

    (5) heat and light

Check your answers on pages 225–226.

CHEMISTRY

Many chemical reactions take place to produce the bright lights of fireworks.

The study of matter and how it changes is called **chemistry**. The alchemists of the 1200s were the first people to study chemistry. They were mainly concerned with trying to change metals into gold. They also tried to make synthetic gems. In their experiments, the alchemists learned about many properties of matter. As you study chemistry, you will better understand the world around you.

One aspect of chemistry is concerned with the physical properties of matter. What are the building blocks of matter? How does matter change from a solid to a liquid? How do substances dissolve? Answers to these questions help us understand ordinary things like cooking and heating.

Another aspect of chemistry is the changes in energy during chemical reactions. Reactions, such as burning, release energy. Learning about such reactions helps us understand many of the machines around us. It also helps us recognize different sources of air pollution.

The study of radioactivity and nuclear reactions is another branch of chemistry. Nuclear reactions are used to produce electrical energy.

Chemists are scientists who study chemistry. Stanley Lloyd Miller is a chemist who is interested in the origins of life on Earth. He believes that Earth was once very different from how it is now. Oceans covered the planet. The atmosphere was a mix of hydrogen, ammonia, and methane. There were lightning storms.

In 1954, Miller set up an experiment to copy these conditions. He started with pure water and added an atmosphere of hydrogen, ammonia, and methane. This mix was circulated through a device he built. An electric spark imitated the energy from lightning. The experiment ran for a week. Miller then analyzed what was in the water. He found simple organic compounds. These are the building blocks of living things on Earth.

Recently, some scientists have suggested other origins of life on Earth. Some say life came from another planet. Others say life first came from deep vents in the ocean floor. These hypotheses are based on computer models. They have not been tested in the lab.

Today Miller is a professor of chemistry at the University of California. He is skeptical about these new ideas, and believes that his hypothesis is correct. He says that test-tube experiments beat computer tests.

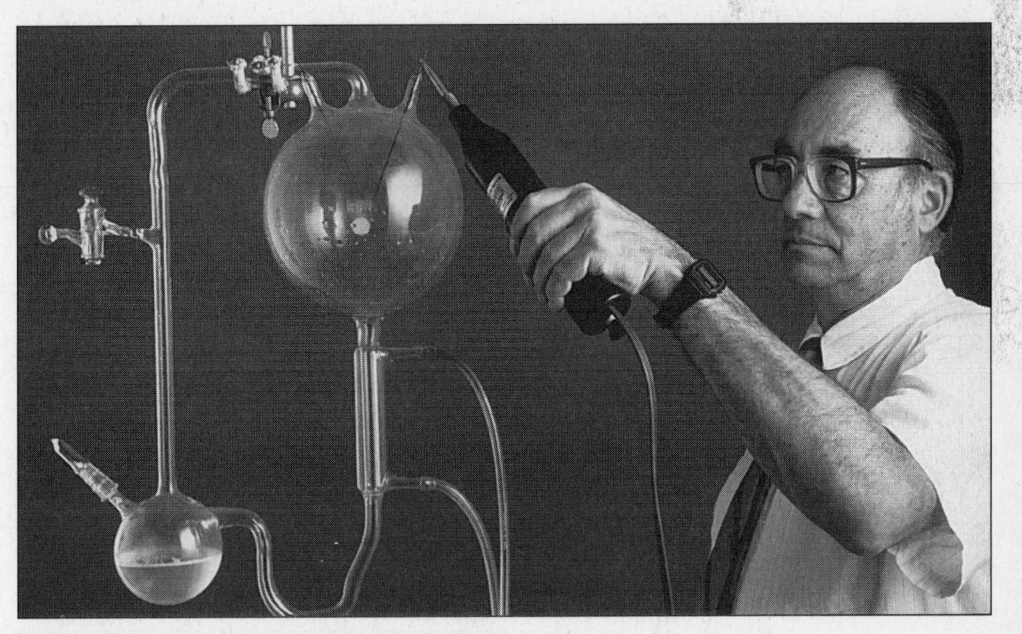

Stanley Miller and a replica of his experiment

This unit features articles about several aspects of chemistry.

- Articles about matter show the chemical nature of household cleaners. They explain why stirring and heating make solids dissolve faster.
- Articles about changes in matter explain what happens when you cook a hamburger and how combustion produces air pollution.
- An article about nuclear reactions shows what radioactivity and chain reactions are.

Matter

Setting the Stage

We use chemistry all the time without even being aware of it. Even doing laundry involves chemistry. Chemists know about matter and how it behaves. Chemists have developed a shorthand to describe different types of matter and chemical changes.

Past: What you already know

You may already know something about matter or chemical changes. Write three things you already know.

1. ______________________________

2. ______________________________

3. ______________________________

Present: What you learn by previewing

Write the headings from the article on pages 137–139 below.

The Chemistry of Cleaning

4. ______________________________

5. ______________________________

6. ______________________________

What does the diagram on page 137 show?

7. ______________________________

Future: Questions to answer

Write three questions you expect this article to answer.

8. ______________________________

9. ______________________________

10. ______________________________

 Check your answers on page 226.

The Chemistry of Cleaning

As you read each section, circle the words you don't know. Look up the meanings.

Almost everyone has run into the laundry problems of ring around the collar or stubborn yellow stains. Makers of detergents and bleaches claim their products can remove the toughest stains. Detergents can remove many stains. But recently scientists figured out why some oily yellow stains won't go away.

If the clothing is washed right away, the oily stain can be removed. But what happens if the stain sits for a week? The aging oil can combine with oxygen from the air. This process changes the colorless oil to a yellow substance. The yellow substance reacts with the fabric. In effect the clothing is dyed yellow.

Substances and Mixtures

Solving the ring-around-the-collar mystery is just one practical application of chemistry. You use chemistry whenever you clean the house or do the laundry. Knowing about the substances and mixtures you use helps you understand this chemistry.

Everything in the world is made up of matter. Matter can be divided into two groups—mixtures and substances. A **mixture** is a combination of two or more kinds of matter that can be separated by physical means. All of the matter in a **substance** is the same.

There are two kinds of substances. An **element** is a substance that cannot be broken down into other substances by ordinary means such as heating or crushing. Two or more elements can combine chemically to form a **compound**. The smallest particle of an element is an **atom**. The smallest particle of a compound is a **molecule**.

All matter is either a mixture or a pure substance.

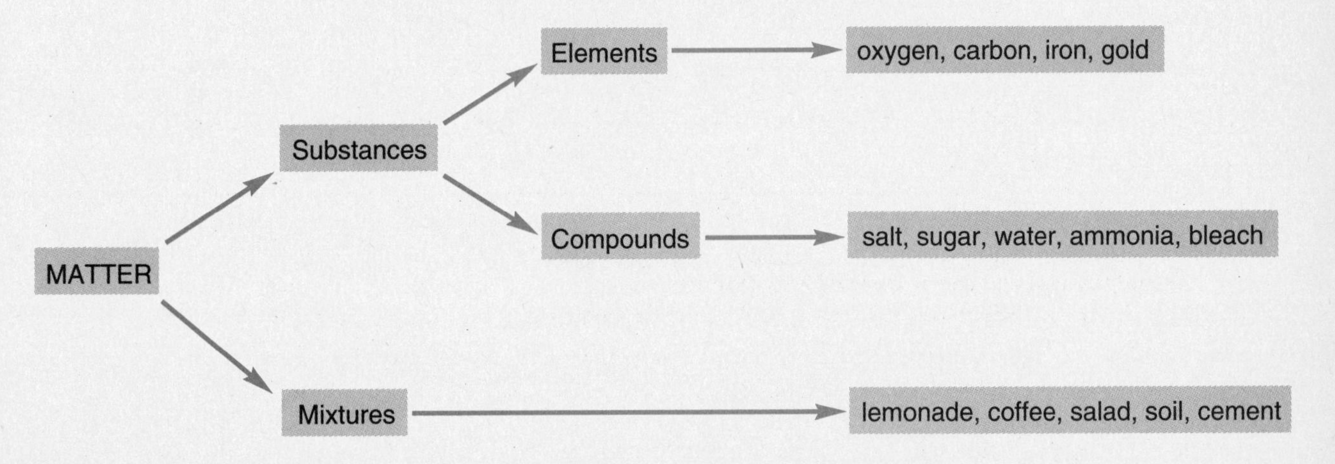

Symbols of Common Elements

Element	Symbol
Hydrogen	H
Carbon	C
Nitrogen	N
Oxygen	O
Sodium	Na
Magnesium	Mg
Sulfur	S
Chlorine	Cl
Iron	Fe
Calcium	Ca

Chemical Symbols and Formulas

When writing about matter and its changes, chemists use a kind of shorthand. A **chemical symbol** of one or two letters stands for each element. The symbols for some common elements are shown in the table on this page. When elements combine to form a compound, the symbols are grouped together in a **chemical formula**. For example, H_2O is the formula for water. The formula shows that the elements hydrogen and oxygen make up the compound water. The formula also shows there are two atoms of hydrogen for each atom of oxygen in the compound. The formula for ammonia is NH_3.

Applying Your Skills and Strategies

Understanding Chemical Formulas. Chemical formulas can tell you a great deal if you know how to decode them. They tell what elements are in a compound. Formulas also show how many atoms of each element make up each molecule of the compound. For example, the formula for sugar is $C_{12}H_{22}O_{11}$. This stands for 12 atoms of carbon, 22 atoms of hydrogen, and 11 atoms of oxygen.

Reread the paragraph under the heading *Chemical Symbols and Formulas* at the top of this page. Name the elements in the compound ammonia.

___

How many atoms of each element are in the compound ammonia?

___

Chemical Reactions

Elements or compounds are changed into other substances in a **chemical reaction**. Iron is a gray solid. Oxygen is a colorless gas. When iron and oxygen combine, they form a new substance. This substance is iron oxide, or rust. It is a brownish red or orange solid. In a reaction, the substances that react are called the **reactants**. In this example, iron and oxygen are the reactants. The substances that form in a reaction are called the **products**. In this example, there is only one product, iron oxide.

Scientists use equations to describe reactions. A **chemical equation** shows the reactants and products of a reaction. This equation shows how rust forms.

iron + oxygen ⟶ iron oxide

Rust stains are hard to remove from clothing. You can remove the stain if you reverse the reaction. The acid in lemon juice can remove the oxygen from iron oxide. The remaining iron atoms rinse away.

Check your answers on page 226.

Compounds in Household Cleaners

Another example of a chemical reaction is the effect of bleach on clothing. The formula for chlorine bleach is NaOCl. In water, this compound produces salt and oxygen. The release of oxygen causes the whitening of the fabric. Nonchlorine bleaches use other sources of oxygen. One source is hydrogen peroxide, H_2O_2. Another source is calcium carbonate, $CaCO_3$.

Applying Your Skills and Strategies

Roots, Suffixes, and Prefixes. The name of a compound tells what's in it. Sometimes the suffix *-ide* is added to the root name of the second element of a compound. This means the compound consists of two elements. Table salt, NaCl, consists of sodium and chlorine but is named sodium chlor*ide*. If the second element has oxygen with it, the suffix *-ate* is added. For example, calcium carbon*ate* is $CaCO_3$.

What elements are in the compound barium chloride?

What elements are in the compound calcium chlorate?

Prefixes are used when more than one compound can be made from the same elements. For example, CO and CO_2 are both made from carbon and oxygen. Prefixes are used to distinguish the two compounds. So CO is carbon monoxide and CO_2 is carbon dioxide. Some common prefixes are shown in the table at the right. How many chlorine atoms are in the compound carbon tetrachloride?

Prefix	Meaning
mono-	one
di-	two
tri-	three
tetra-	four

What is the name of the compound that has one aluminum atom and three oxygen atoms?

Check your answers on page 226.

Thinking About the Article

Fill in the blank with the word or words that best complete each statement.

1. ________________ are combinations of two or more kinds of matter that can be separated by physical means.
2. All of the matter in a ________________, such as water or oxygen, is the same.
3. ________________ are substances that cannot be made into other substances by ordinary means, such as heating or crushing.
4. Chemical ______________ are one or two letters used to represent elements.
5. A chemical ______________ is a group of symbols that shows the makeup of a compound.
6. When elements or compounds are changed into one or more different substances, a chemical ________________ takes place.
7. A chemical ________________ shows the reactants and products in a chemical reaction.

Identify each of the following as an element or a compound.

8. O_2 ____________________
9. CO ____________________
10. NaCl ____________________
11. NaOH ____________________

Write your answers in the space provided.

12. Review the questions you wrote on page 136. Did the article answer your questions? If you said *yes*, write the answers. If your questions were not answered, write three things you learned from this article.

____

____

____

____

____

Check your answers on page 226.

13. What is the chemical symbol for sodium?

____

Circle the number of the best answer.

14. Which of the following is an example of a mixture?
    (1) water
    (2) salt (sodium chloride)
    (3) carbon dioxide
    (4) gold
    (5) lemonade

15. What is the name of the compound MgS?
    (1) magnesium sulfur
    (2) magnesium sulfide
    (3) magnesium disulfide
    (4) sulfur magnesiate
    (5) sulfide magnesium

16. What are the reactants in the equation?

    iron + chlorine ⟶ iron chloride

    (1) iron chloride only
    (2) iron only
    (3) chlorine only
    (4) iron and iron chloride
    (5) iron and chlorine

Write your answer in the space provided.

17. Name three common household cleaning compounds and describe how you or someone you know uses them.

____

____

____

____

____

Check your answers on pages 226–227.

Changes in Matter

Setting the Stage

All matter on Earth is either a solid, a liquid, or a gas. You often change matter from one state to another when you cook. Cooking also involves chemical reactions in the food.

Past: What you already know

You may already know something about changing matter from one state to another. Write three things you already know.

1. ______________________________

2. ______________________________

3. ______________________________

Present: What you learn by previewing

Write the headings from the article on pages 143–145 below.

Cooking with Chemistry

4. ______________________________

5. ______________________________

6. ______________________________

What do the photos on page 143 show?

7. ______________________________

____

Future: Questions to answer

Write three questions you expect this article to answer.

8. ______________________________

9. ______________________________

10. ______________________________

Check your answers on page 227.

Cooking with Chemistry

As you read each section, circle the words you don't know. Look up the meanings.

Ice cubes, water, and steam are used by cooks. Ice cubes are used to chill liquids. Water is used to boil food and as an ingredient in many recipes. Steam, which is actually water vapor, is used to cook vegetables. What do ice, water, and steam have in common? They are three forms, or states, of the same compound, H_2O.

Each state of matter has its own properties, or characteristics. A **solid** has a definite shape and takes up a definite amount of space. A **liquid** takes up a definite amount of space, but it doesn't have a definite shape. A liquid flows and takes the shape of its container. A **gas** does not have a definite size or shape. It expands to fill its container. If you remove the lid from a pot of steaming vegetables, water vapor escapes into the kitchen.

Changes of State

Matter changes state when energy, in the form of heat, is added to or removed from a substance. When you leave an ice cube tray on the counter, the ice absorbs heat from the air. Eventually the ice cubes melt. **Melting** is the change from a solid to a liquid. If you put the tray back in the freezer, the water will change back into ice. **Freezing** is the change from a liquid to a solid.

When you heat water, bubbles of gas form. They rise and burst on the surface of the water. **Boiling** is the rapid change from a liquid to a gas. A liquid can also change to a gas slowly through **evaporation**. If a glass of water is left out for a long time, the water evaporates from its surface. The reverse of boiling or evaporation is **condensation**. This is the change from a gas to a liquid. If a cold soda bottle is taken from the refrigerator, water vapor in the air will condense on the surface of the bottle.

Melting and boiling are two changes in state.

Physical Changes and Chemical Changes

A **physical change** is one in which the appearance of matter changes but its make-up and most of its properties stay the same. The boiling of water, the melting of butter, the dissolving of sugar in tea, and the smashing of a plate are physical changes. No new matter is formed. Matter is changed from one state to another in boiling and melting. When sugar dissolves, two kinds of matter are mixed. When a plate breaks, it changes size and shape.

Unlike a physical change, a **chemical change** makes new substances. Making a lemon fizz with baking soda and lemonade involves a chemical change. Baking soda mixed with an acid, like lemonade, gives off the gas carbon dioxide. This is a new product.

Many activities involve physical and chemical changes. Making an omelet is an example. Breaking and scrambling the eggs are physical changes. The chemical make-up of the eggs has not changed. You have just mixed the parts together. Cooking the eggs is a chemical change. The heat makes the proteins in the egg harden. The make-up of the cooked eggs is different from the make-up of the raw eggs.

Applying Your Skills and Strategies

Finding the Main Idea. The *main idea* of a paragraph tells what the paragraph is about. The sentences of the paragraph contribute *details* to the main idea. The main idea of the first paragraph on this page is that no new matter forms in a physical change.

What is the main idea of the second paragraph on this page?

____

____

Cooking a Hamburger

Cooking and eating a hamburger involves physical and chemical changes. When the butcher grinds beef to make hamburger meat, the meat is ground into small pieces. No new substances are made, so grinding beef is a physical change.

Until the meat is packaged in plastic, the surface of the meat reacts with oxygen in the air. Myoglobin, a chemical in the beef, combines with oxygen. This chemical change is called **oxidation**. This change turns the surface of the meat bright red.

The next step is to form hamburger patties. In this step, you are changing the shape of the ground beef. There aren't any chemical changes, just a physical change.

Check your answer on page 227.

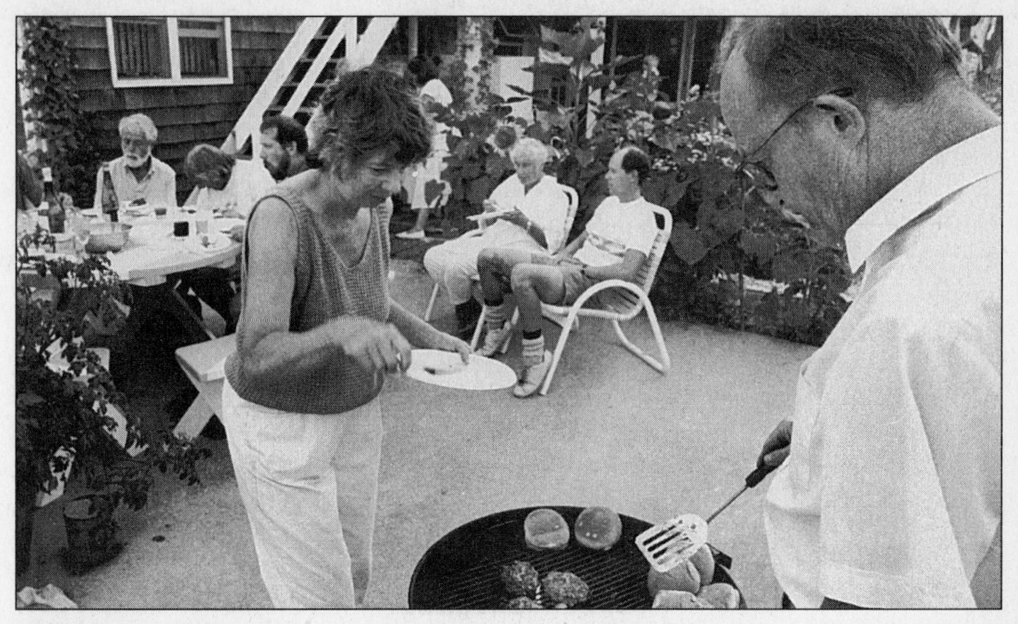

Changes in a hamburger

Now the hamburgers are ready for cooking. Many cooks quickly sear one side, then the other. Searing causes a chemical change in the surface proteins, and they form a crust. The crust keeps too much water from evaporating. It also keeps some of the fat from melting and seeping out of the hamburger. Both of these physical changes would make the hamburger dry.

Applying Your Skills and Strategies

Applying Knowledge to Other Contexts. Information becomes more valuable when you use it. This article describes changes in a hamburger as it is prepared and cooked. Apply this knowledge to other foods. Label each of the following as a *chemical* or a *physical* change.

A piece of apple turns brown when it is exposed to air.

____

A spoonful of sugar is dissolved in a cup of coffee.

____

As the hamburger cooks, the inside loses its red color. This is caused by another chemical change in myoglobin. At the same time, chemical changes in the protein make the meat become firmer.

When you eat the hamburger, more physical and chemical changes occur. Your teeth cut and grind the hamburger. This is a physical change. The hamburger is broken down chemically by substances in your digestive system. Digestion is another series of physical and chemical changes.

Check your answers on page 227.

Thinking About the Article

Fill in the blank with the word or words that best complete each statement.

1. A ____________ is a state of matter that has a definite shape and takes up a definite amount of space.

2. A ____________ is a state of matter that takes up a definite amount of space but doesn't have a definite shape.

3. A ____________ is a state of matter that does not have a definite shape or size.

4. A __________________ change is one in which the appearance of matter changes, but its make-up and most of its properties remain the same.

5. A _______________ change makes new substances not present before the change.

6. A liquid slowly changes to a gas through ______________________.

7. _______________ is the change from a liquid to a solid.

Identify each of the following as a chemical or physical change.

8. Melting butter ____________________________.

9. Chopping onions ____________________________.

10. Hard-boiling an egg ____________________________.

Write your answer in the space provided.

11. Review the questions you wrote on page 142. Did the article answer your questions? If you said *yes,* write the answers. If your questions were not answered, write three things you learned from this article.

____

____

____

____

____

Check your answers on page 227.

Circle the number of the best answer.

12. Which title best describes this article?

    (1) Chemical Changes

    (2) Physical Changes

    (3) Chemical Changes in Cooking

    (4) Chemical and Physical Changes in Cooking

    (5) Cooking a Hamburger

13. The bubbles in a glass of soda rise to the top. If the soda is left uncovered, it eventually loses its fizz. This is an example of

    (1) melting.

    (2) condensation.

    (3) a chemical change.

    (4) a physical change.

    (5) oxidation.

14. All of the following are physical changes except

    (1) pouring orange juice into a glass.

    (2) freezing a popsicle.

    (3) mixing a milk shake.

    (4) cooking an egg.

    (5) bringing soup to a boil.

Write your answers in the space provided.

15. List three examples of physical changes not mentioned in this article.

____

____

____

16. Describe an experience you or someone you know has had while cooking. Identify any physical or chemical changes that occurred in the food as it was cooked.

____

____

____

Check your answers on page 227.

Solutions

Setting the Stage

Most types of matter are mixtures of two or more substances. A solution is one type of mixture. It has the same makeup throughout. We use many solutions in our daily lives. Soda, brass, gasoline, and a cup a coffee are examples of solutions.

Past: What you already know

You may already know something about mixtures or solutions. Write three things you already know.

1. ______________________________

2. ______________________________

3. ______________________________

Present: What you learn by previewing

Write the headings from the article on pages 149–151 below.

Mixing It Up

4. ______________________________

5. ______________________________

6. ______________________________

What do the photos on page 149 show?

7. ______________________________

Future: Questions to answer

Write three questions you expect this article to answer.

8. ______________________________

9. ______________________________

10. ______________________________

Check your answers on page 227.

Mixing It Up

As you read each section, circle the words you don't know. Look up the meanings.

Mixtures are all around you. The paper in this book is a mixture of fibers. The inks with which it is printed are mixtures of colored substances. Even the air around you is a mixture. It contains nitrogen, oxygen, carbon dioxide, and other gases.

What is a mixture? A **mixture** is made up of two or more substances that can be separated by physical means. The properties of a mixture are the properties of its ingredients. That is why sugar water is sweet and wet. Neither the sugar nor the water loses its properties when they are mixed together. Another characteristic of a mixture is that the amounts of its ingredients can vary.

Mixtures can be solids, liquids, or gases. A nickel is a solid mixture of the elements nickel and copper. Blood is a liquid mixture of water, cells, proteins, oxygen, carbon dioxide, sugar, and other substances. Soda is a mixture of flavored liquid and the gas carbon dioxide.

The substances in a mixture can be separated by physical means. A mixture of red and blue blocks can be sorted by hand. A filter can be used to separate sand from water. A magnet can separate a mixture of steel and plastic paper clips. Liquid mixtures can be separated by **distillation**. Since liquids boil at different temperatures, they can be condensed and collected as they boil out of the mixture. That's how alcohol is obtained from fermented juices and brewed grains.

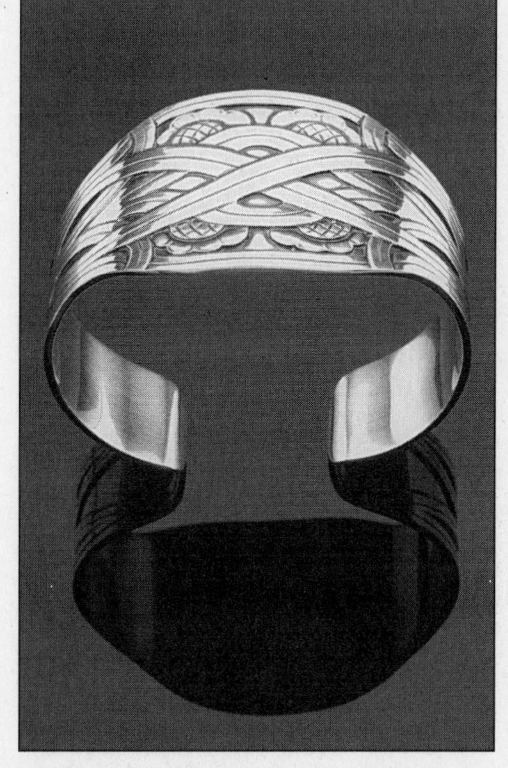

Sterling silver is a mixture of silver and copper.

Milk is a mixture of water, proteins, fat, and sugar.

What Is a Solution?

A solution is a type of mixture. In a **solution** the ingredients are distributed evenly throughout. All samples taken from a solution have the same amount of each substance. That's why the first and last sip from a can of soda taste the same. Solutions can be made of solids, liquids, or gases. Brass is a solid solution. Tea is a liquid solution. Seltzer is a solution of carbon dioxide gas in water.

In a solution, the substance that is present in the greater amount is called the **solvent**. Water is the most common solvent. The substance present in the smaller amount is called the **solute**. The solvent and solute may be in different states before the solution is formed. However, the final state of the solution will be that of the solvent. So a solution of water and powdered fruit drink is a liquid, not a solid.

Applying Your Skills and Strategies

Comparing and Contrasting. When learning about things that are related, it is helpful to compare and contrast them. Comparing is pointing out how two things are alike. Contrasting is showing how two things are different. This article tells several things about mixtures and solutions. How are mixtures and solutions alike?

____

____

How are mixtures and solutions different?

____

____

How Solutions Form

When a solute dissolves, its particles spread evenly throughout the solution. How quickly the solute dissolves depends on several things. The smaller the particles of solute, the more quickly they dissolve. That's why instant coffee is made of small grains, not large chunks.

Stirring or shaking make a solute dissolve faster. The movement brings the solvent in contact with more of the solute. Stirring a cup of instant coffee makes the coffee dissolve more quickly.

Heat also makes a solute dissolve faster. Molecules move more quickly when they are hot. You can make instant coffee more quickly with boiling water than with cold water.

Check your answers on page 228.

Solubility

The amount of a solute that will dissolve in a given amount of solvent at a given temperature is called its **solubility**. The effect of temperature on solubility is shown in the graph at the bottom of this page. Each line on the graph is called a solubility curve. You can use this graph to find the solubility of substances in water at any temperature.

Applying Your Skills and Strategies

Reading a Line Graph. Like other graphs, line graphs show information instead of presenting it in words. The title tells you what the graph shows. Circle the title of the graph on this page. The labels along the side and bottom of the graph tell how the graph is organized. What do the labels on the side of the graph show?

____

What do the labels along the bottom of the graph show?

____

Now look at the curve for sodium nitrate. How many ounces will dissolve in one quart of boiling water (212°F)?

____

Solubility of Some Solids in Water

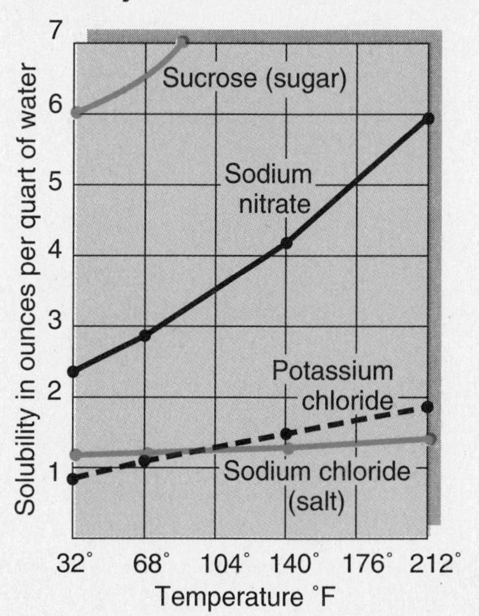

The solubility of solids and liquids usually increases as the temperature rises. However, the opposite is true of gases. As the temperature rises, dissolved gas particles gain energy. They escape from the surface of the solution. That's why an opened bottle of soda goes flat more quickly at room temperature than in the refrigerator.

What happens if you add more solute than the solvent can hold? Extra solute settles to the bottom of the mixture. That is why there is a limit as to how sweet you can make iced tea. Once you reach the limit of solubility of the sugar, no more dissolves. The sugar on the bottom of the glass doesn't make the tea sweeter.

If the temperature of a sugar solution changes, the solubility of the sugar changes. Suppose you dissolve all the sugar you can in hot water. Then you let the water cool. Sugar crystals will come out of the solution. If you let the water evaporate, the sugar crystals will be left behind. This is how rock candy is made.

Check your answers on page 228.

Thinking About the Article

Fill in the blank with the word or words that best complete each statement.

1. A ______________ is made up of two or more substances that can be separated by physical means.
2. ________________________ is a process of boiling and condensing that is used to separate liquids in a mixture.
3. A mixture in which the substances are distributed evenly throughout is called a ________________.
4. The substance present in the greater amount in a solution is called the ______________.
5. The substance present in the smaller amount in a solution is called the ____________.

Identify each of the following as a mixture or solution.

6. Coffee ________________
7. Chocolate chip cookie ________________
8. Milk ________________

Write your answer in the space provided.

9. Review the questions you wrote on page 148. Did the article answer your questions? If you said *yes*, write the answers. If your questions were not answered, write three things you learned from this article.

____

____

____

____

____

 Check your answers on page 228.

Circle the number of the best answer.

10. Which of the following is an example of distillation?
    (1) separating alcohol from brewed grain
    (2) creating salt by combining sodium and chlorine
    (3) allowing a glass of soda to go flat
    (4) filtering sand out of a sand and water mixture
    (5) putting a spoonful of sugar in a cup of coffee

11. Which of the following solutions is a solid at room temperature?
    (1) salt dissolved in water
    (2) water vapor dissolved in air
    (3) zinc dissolved in copper
    (4) alcohol dissolved in water
    (5) water dissolved in alcohol

12. Approximately how much sugar will dissolve in 1 quart of water at 33°F?
    (1) 1 ounce
    (2) 2 ounces
    (3) 4 ounces
    (4) 6 ounces
    (5) 7 ounces

Write your answers in the space provided.

13. Some Middle Eastern countries get a great deal of their drinking water from the ocean. What process could be used to purify the water?

____

____

14. Describe an experience you or someone you know has had with a mixture or solution.

____

____

____

Check your answers on page 228.

Combustion

Setting the Stage

Many devices, such as kerosene heaters, use a chemical reaction to produce heat. Besides heat, they also give off some indoor air pollutants.

Past: What you already know

You may already know something about heaters, chemical reactions, or air pollution. Write three things you already know.

1. ______________________________

2. ______________________________

3. ______________________________

Present: What you learn by previewing

Write the headings from the article on pages 155–157 below.

The By-Products of Burning

4. ______________________________

5. ______________________________

6. ______________________________

7. ______________________________

What does the photo on page 155 show?

8. ______________________________

____

Future: Questions to answer

Write two questions you expect this article to answer.

9. ______________________________

10. ______________________________

 Check your answers on page 228.

The By-Products of Burning

As you read each section, circle the words you don't know. Look up the meanings.

Humans first used fire in prehistoric times. Since then, people have been burning fuels to produce heat. At first, people burned wood for warmth. During the 1700s, the first engines were invented. In an engine, a fuel is burned and heat energy is captured. This energy is then used to produce motion. Today, people burn fuels in engines, cooking stoves, and heating devices.

Combustion Reactions

Burning is a chemical change that chemists call **combustion**. In combustion, oxygen reacts with a fuel. Heat and light energy are released. An example of combustion is the burning of a fuel, such as kerosene, oil, or gas. These fuels are **hydrocarbons**, or compounds made only of hydrogen and carbon. When they burn, the hydrogen and carbon combine with oxygen from the air. Carbon dioxide and water vapor are produced.

Combustion reactions need a little energy to get them started. This energy is called **activation energy**. To start a twig burning, you must light a match to it. Once combustion starts, no additional energy is needed.

Each substance has its own **kindling temperature**. That's the temperature to which a substance must be heated before it will burn. The form of the substance affects the kindling temperature. For example, sawdust catches fire faster than a log. Vaporized gasoline ignites more easily than liquid gasoline.

The burning of gas in this motorcycle engine is an example of combustion.

How a Combustion Heater Works

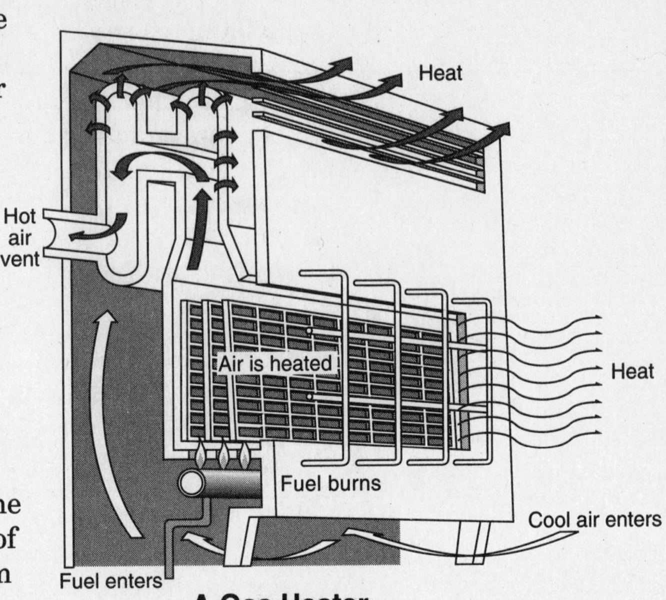

A Gas Heater

A small kerosene or natural gas heater can be used to warm one or two rooms. A kerosene heater has a small built-in fuel tank. Some natural gas heaters have fuel tanks, too. Other gas heaters are connected to the gas lines of the house. The fuel is piped to the burners. Most heaters use an electric spark to start the combustion reaction. Cool air from the room enters at the base of the heater. The heat from combustion warms the air. Warmed air flows into the room. In a gas heater, the waste products of combustion leave the house through a vent. In a kerosene heater, the hot air flows into the room.

Kerosene heaters have many advantages. A kerosene heater is not very expensive. It does not have to be attached to a chimney, so it can be moved from room to room. These heaters warm up quickly and provide heat for up to thirty hours without refueling.

Applying Your Skills and Strategies

Making Predictions. Active readers think about what they have read. They make predictions about what will come next in the article. Reread the two paragraphs above. What do you predict the article will discuss next?

____

____

____

Kerosene heaters also have some disadvantages. If not used properly, they can be unsafe. If the heater is close to drapes, they may catch fire. The outside of the heater is very hot and can cause burns. The heater may tip over, causing injury or a fire. These are all obvious hazards. Less obvious is the hazard caused by the release of pollutants into the air.

Check your answer on page 228.

Indoor Air Pollution

Complete combustion produces carbon dioxide and water vapor. Combustion is complete only if the fuel is pure and there is plenty of oxygen. Often the fuel is not pure. Also, there may be too little oxygen. Then combustion is incomplete. Other substances are released along with the usual products of combustion. These substances pollute the air.

Applying Your Skills and Strategies

Understanding the Implied Main Idea. Sometimes a writer does not actually state the main idea of a paragraph. You have to "read between the lines" and add up the details to figure out the main idea. The paragraph above describes complete combustion and incomplete combustion. Write the main idea in your own words.

____

____

____

If the room is airtight, there can be a high level of pollutants in the air. The air may contain carbon monoxide, nitric oxide, nitrogen dioxide, or sulphur dioxide. These substances can harm the eyes, throat, and lungs. Carbon monoxide is especially dangerous. You can't smell it or see it. Yet high levels of carbon monoxide can cause death.

Both natural gas and kerosene heaters can give off pollutants. Natural gas burns more cleanly than kerosene. But makers of kerosene heaters claim that these heaters can burn cleanly, too. If high-quality kerosene is used, the combustion reaction is 99.5 percent complete. This means that only 0.5 percent of the reaction is incomplete.

Kerosene and natural gas heaters are not the only producers of indoor air pollution. Wood stoves, fireplaces, gas stoves, and tobacco smoke are other sources. Leaking chimneys and furnaces also produce indoor air pollution.

Safety Precautions

People using kerosene heaters or other types of combustion devices indoors should always follow the manufacturer's instructions. The heater should always be in a safe place. This means it should be placed away from materials that might burn and away from places where children play.

To reduce indoor pollution, use only high-quality kerosene. Keep the doors to other rooms open for ventilation. Remember that combustion uses oxygen from the air. If the house is airtight, it may be necessary to open a window slightly to let in more oxygen.

Check your answer on page 228.

Thinking About the Article

Fill in the blank with the word or words that best complete each statement.

1. Burning is called ____________________.
2. A ______________________ is a compound made only of hydrogen and carbon.
3. The energy required to start a combustion reaction is called

   ____________________ energy.
4. The temperature to which a substance must be heated before it will

   burn is called its ________________ temperature.

Write your answers in the space provided.

5. Review the questions you wrote on page 154. Did the article answer your questions? If you said *yes*, write the answers. If your questions were not answered, write three things you learned from this article.

____

____

____

____

____

6. What gas from the air is needed for combustion?

____

7. Describe what happens during a combustion reaction.

____

____

8. What conditions are needed for combustion to be complete?

____

____

____

Check your answers on page 228.

Circle the number of the best answer.

9. Which of the following is not a combustion device?
   (1) blowtorch
   (2) kerosene heater
   (3) solar battery
   (4) gas stove
   (5) car

10. What are the products of a complete combustion reaction involving hydrocarbons and oxygen?
   (1) carbon dioxide and water vapor
   (2) carbon monoxide and sodium chloride
   (3) carbon monoxide and water vapor
   (4) nitrogen oxide and carbon dioxide
   (5) carbon dioxide and sulfur dioxide

11. Which pollutant is released by a kerosene heater?
   (1) acid rain
   (2) carbon dioxide
   (3) water vapor
   (4) carbon monoxide
   (5) smog

Write your answers in the space provided.

12. How can some of the hazards of using a kerosene heater or other combustion device be reduced?

____

____

____

13. Describe a combustion device in your home. What type of fuel is used? Do you think it is producing any indoor air pollutants?

____

____

____

Check your answers on page 229.

Nuclear Fission

Setting the Stage

Nuclear reactors produce large amounts of energy with small amounts of fuel. About 10 percent of the electricity generated in the United States comes from nuclear reactors.

Past: What you already know

You may already know something about nuclear power or generating electricity. Write three things you already know.

1. ______________________________

2. ______________________________

3. ______________________________

Present: What you learn by previewing

Write the headings from the article on pages 161–163 below.

Using Nuclear Power

4. ______________________________

5. ______________________________

6. ______________________________

What does the diagram on page 163 show?

7. ______________________________

____

Future: Questions to answer

Write three questions you expect this article to answer.

8. ______________________________

9. ______________________________

10. ______________________________

Check your answers on page 229.

Using Nuclear Power

As you read each section, circle the words you don't know. Look up the meanings.

There's something eerie about a visit to a nuclear power plant. The many power lines let you know that huge amounts of electricity are being sent out. Yet, oddly enough, the power plant is quiet. You expect a power plant to give off smoke. But there aren't any smokestacks. What goes on inside this silent giant?

A Modern Energy Source

Early in this century, scientists predicted that the atom could produce huge amounts of energy. But this wasn't proved until 1939. In that year, a group of scientists split the atom for the first time. Atoms must be split to produce nuclear energy. Since then, people have been finding different uses for nuclear energy.

Nuclear energy comes from inside an atom. The reactions that release this energy are more powerful than ordinary chemical reactions. As a result, nuclear power plants produce a lot of electricity from very little fuel. Some countries, such as France, depend on nuclear power for more than half of their electricity. In the United States, about 10 percent of the electricity comes from nuclear power plants.

A nuclear power plant

Nuclear Reactions

Nuclear reactions are changes in the nucleus, or center, of an atom. The nucleus of an atom contains two kinds of particles. **Protons** are positively charged particles. All atoms of a substance have the same number of protons. **Neutrons** are particles that have no charge. The number of neutrons in a nucleus can vary.

Many atoms have about the same number of protons and neutrons. Some atoms have an unbalanced number of these particles. This causes the nucleus to be unstable. An unstable nucleus gives off particles until it becomes stable. This change is called **radioactive decay**. When an atom decays, it becomes another substance. For example, uranium decays to form lead. This is a natural nuclear reaction. Atoms that decay are **radioactive**. They give off particles and energy in the form of radiation.

Applying Your Skills and Strategies

Using a Glossary or Dictionary. The glossary at the end of this book is a good source of information. If you don't understand a word, look it up in the glossary. If the word is not contained in the glossary, look it up in a dictionary. Go back over the first two paragraphs on this page. Circle any words you do not know. Look up their meanings in the glossary or a dictionary. Then write their meanings on a separate sheet of paper.

Some radioactive atoms can be split. The splitting of an atom's nucleus is called **fission**. This reaction releases both energy and radiation. It also gives off more neutrons. Fission does not occur in nature. It occurs when neutrons are shot at an unstable nucleus. The nucleus splits, and two new nuclei are formed.

The neutrons released during one fission strike other atoms, producing more fissions. This is called a chain reaction. A **chain reaction** is one that keeps itself going. Toppling a row of dominoes by knocking over the first one is similar to a chain reaction.

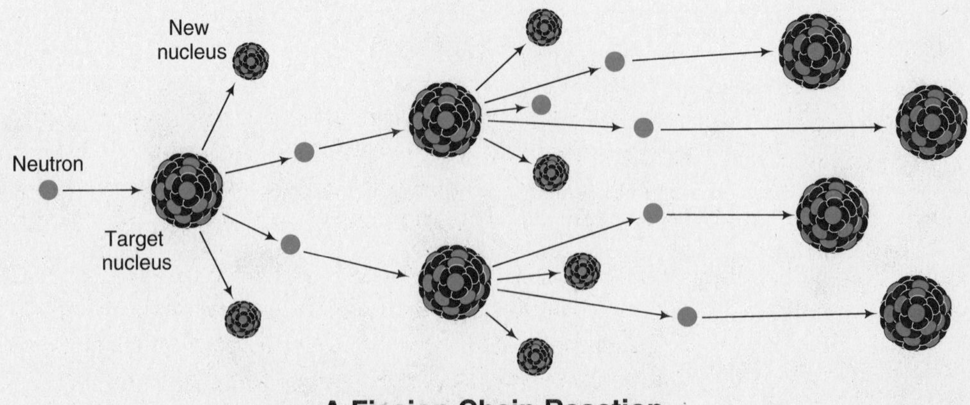

A Fission Chain Reaction

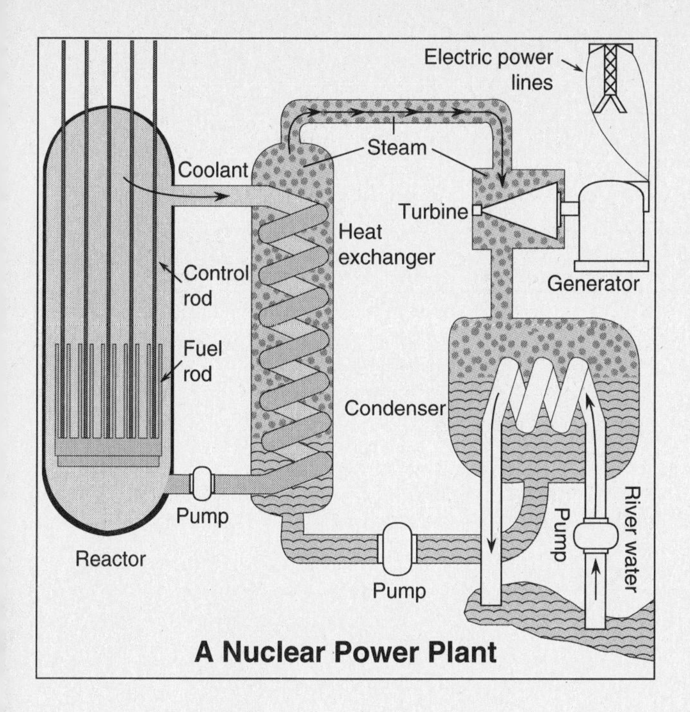

A Nuclear Power Plant

How a Nuclear Power Plant Works

The fission reactions at a nuclear power plant occur in a structure called a reactor. The core of the reactor holds the fuel rods and control rods. The fuel rods contain plutonium. The control rods keep the chain reaction at a steady rate. These rods are made of a substance that absorbs neutrons. If the reaction is too slow, the control rods are pulled out a bit. This makes more neutrons available to split more atoms. The control rods are pushed back in if the reaction speeds up too much.

The fission reaction gives off heat. A coolant, usually water, absorbs this heat. The coolant carries the heat away from the core into a heat exchanger. There the heat from the coolant is used to boil water. The steam turns a turbine, which generates electricity.

Applying Your Skills and Strategies

Sequencing. The order in which things happen is very important. If the steps in a process such as generating electricity are out of sequence, the process doesn't work. Review the diagram and the first two paragraphs under the heading *How a Nuclear Power Plant Works.* Describe the process of generating electricity from a nuclear reaction in five steps.

1. ______________________________

2. ______________________________

3. ______________________________

4. ______________________________

5. ______________________________

The reactor is housed in a containment building, which has thick concrete walls that can absorb radiation. If the core gets too hot, safety systems provide emergency cooling. The reaction can be shut down quickly by pushing the control rods all the way into the core.

Nuclear power plants are designed to be as safe as possible. However, accidents do happen. In 1979, a reactor in Pennsylvania overheated and had to be shut down. In 1986, a reactor in the Soviet Union caught fire. Some power plants are poorly designed and not built well. Also, people make mistakes. If nuclear plants are to be the power source of the future, they must be made better and safer. Their operators must be trained to handle emergencies as well as routine tasks.

Check your answers on page 229.

Thinking About the Article

Fill in the blank with the word or words that best complete each statement.

1. ______________________ is energy released from inside an atom.
2. Atoms that are ______________________ decay and become another substance.
3. ______________________ are changes in the nucleus of an atom.
4. The positively charged particles in the nucleus of an atom are called ______________.
5. ______________ are particles that are found in the nucleus and have no charge.
6. The splitting of an atom's nucleus into two smaller parts is called ______________.

Write your answers in the space provided.

7. Review the questions you wrote on page 160. Did the article answer your questions? If you said *yes*, write the answers. If your questions were not answered, write three things you learned from this article.

____

____

____

____

____

8. What happens in a chain reaction?

____

____

9. What type of fuel is used in nuclear reactors?

____

Check your answers on page 229.

Circle the number of the best answer.

10. Arrange the following steps in the correct order. Refer to the diagram and text on page 163.

    a. Steam turns the turbine.
    b. The generator produces electricity.
    c. Heat from the coolant boils water.
    d. Fission heats the coolant.

    (1) a, b, c, d

    (2) c, d, a, b

    (3) d, a, b, c

    (4) d, c, b, a

    (5) d, c, a, b

11. According to the diagram on page 163, a nuclear power plant uses river water to

    (1) condense steam.

    (2) cool the core.

    (3) turn the turbine.

    (4) clean the fuel rods.

    (5) clean the control rods.

12. The control rods in a reactor keep the chain reaction at a steady rate by absorbing

    (1) neutrons.

    (2) protons.

    (3) plutonium.

    (4) water.

    (5) atoms.

Write your answer in the space provided.

13. Is any of the electricity in your area produced by a nuclear power plant? If so, what do you and the people in your area think about having a nuclear power plant close by? If not, what would you think about having a nuclear power plant built in your area?

____

____

Check your answers on page 229.

Unit 3 Review:
Chemistry

Mixtures

The *lead* in your pencil is not actually made of the metal lead. It is made mostly of a form of carbon called graphite. Graphite is very soft. As a result, anything you write with graphite will easily smudge. To solve this problem, graphite must be mixed with something else.

A **mixture** is a combination of two or more substances, in which the proportions may vary. A mixture has the properties of the substances in the mixture. These properties vary depending on the make-up of the mixture. For example, sugar is sweet and water is not. A mixture of sugar and water is sweet but not as sweet as sugar alone. If you keep adding water to the mixture, the taste gets less sweet. The property of sweetness is still there, but the amount of sweetness varies.

In a pencil, the *lead* is a mixture of graphite and clay. Graphite is very soft, but clay is hard. The mixture of the two is harder than graphite but softer than clay. You may have noticed a number stamped on a pencil. The number tells how hard the pencil is. Number 2 pencils are fairly soft and are the most common. These pencils make dark lines, but the lines can smudge. As the amount of clay in the mixture increases, the hardness increases. A number 9 pencil is very hard. It makes fine lines that do not smudge.

Fill in the blank with the word or words that best complete each statement.

1. A _______________ is a combination of two or more substances that are combined in varying proportions.

2. The lead in a pencil is a mixture of _________________ and

   _______________.

Circle the number of the best answer.

3. As the amount of clay in a pencil increases,
   (1) the number of the pencil increases.
   (2) the lines it makes become finer.
   (3) the lines it makes smudge less.
   (4) the pencil lead becomes harder.
   (5) all of the above.

Go on to the next page.

Energy and Chemical Reactions

In a **chemical reaction,** elements or compounds are changed into other substances. In this process, energy may be given off or taken in. An **exothermic reaction** gives off energy. Burning, or combustion, is an example of an exothermic reaction. When wood is burned, energy is given off in the form of light and heat.

Photosynthesis is a chemical reaction that takes place in plants. Through this reaction, plants use the energy in sunlight to turn carbon dioxide, minerals, and water into food. This is an example of an **endothermic reaction,** or one that takes in energy.

You may have used an instant hot pack for first aid. These plastic pouches contain chemicals. When you break the seal inside the pouch, the chemicals react. The reaction is exothermic and gives off heat. Once the reaction is finished, no additional heat is given off.

There are also instant cold packs used for first aid. When the chemicals in these pouches react, they do not give off heat. Instead, they take in heat. The reaction is endothermic. Because the reaction absorbs heat, the pouch feels cold when placed against the skin.

Fill in the blank with the word or words that best complete each statement.

4. An ______________________________ gives off energy.

5. An ______________________________ takes in energy.

6. In a ______________________________, elements or compounds are changed into other substances.

7. Burning, or ____________________, is an exothermic reaction that gives off energy in the form of light and heat.

Circle the number of the best answer.

8. Which of the following is an example of an endothermic reaction?

   (1) burning a candle

   (2) combustion of wood

   (3) exploding dynamite

   (4) photosynthesis in a plant

   (5) production of light by a firefly

Go on to the next page.

Putting Out Fires

Fire, or **combustion,** is a useful chemical reaction. However, sometimes a fire gets out of control and must be put out. There are several ways to do this. All methods of putting out a fire work by removing something the reaction needs in order to continue.

One way of putting out a fire is to take away one of the substances that is used in the reaction. The simplest way to do this is to remove the **fuel**, or the material that is burning. You do this when you turn off the gas on the stove. But in a raging fire, it is hard to remove the fuel.

It is easier to remove the oxygen that is needed to keep the reaction going. A small fire can be smothered. Baking soda can be poured on a small grease fire in an oven. The layer of baking soda keeps oxygen away from the grease, which is the fuel. Smothering a campfire with dirt works the same way.

Another way to put out a fire is to take away some of its heat. Materials do not burn until they are heated to the **kindling temperature.** Once a fire is burning, it continues to heat its fuel to the kindling temperature. If you can take enough heat away from the fire, the fuel will be below the kindling temperature and will not burn. This is how water puts out a fire.

A carbon-dioxide fire extinguisher uses two methods at once. The carbon dioxide is heavier than oxygen. It makes a layer below the oxygen but above the fuel. As the carbon dioxide comes out of the extinguisher, it expands rapidly. This process absorbs heat. So the carbon dioxide also cools the burning material.

Circle the number of the best answer.

9. If a fire is to continue burning, it must have enough
   (1) carbon dioxide.
   (2) oxygen and carbon dioxide.
   (3) fuel and heat.
   (4) oxygen and heat.
   (5) oxygen, fuel, and heat.

10. A heavy blanket thrown on a small fire puts out the fire by
    (1) adding carbon dioxide.
    (2) removing oxygen.
    (3) removing fuel.
    (4) removing heat.
    (5) removing carbon dioxide.

Go on to the next page.

Fusion Reactions

Nuclear reactions are changes in the nucleus, or center, of an atom. One kind of nuclear reaction, fusion, is being studied by many scientists. **Nuclear fusion** is the reaction in which two nuclei combine. In the process, the nucleus of a larger atom is formed.

In nuclear fusion, hydrogen nuclei fuse, or join, and form a helium nucleus. A huge amount of energy is released. This reaction takes place only under conditions of great pressure and high temperature. These conditions are found on the sun. Fusion reactions are the source of the energy that the sun gives off.

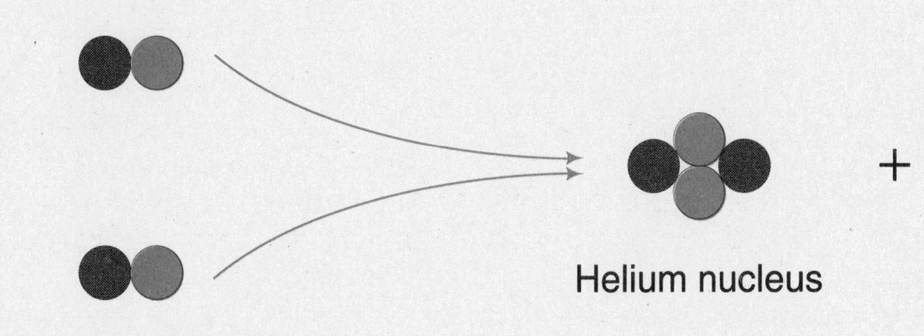

Fusion reactions do not take place naturally on Earth. There isn't a place on the planet as hot as the sun. Scientists are looking for ways to make fusion occur at lower temperatures. If scientists could make such "cold fusion" reactions work, they would have a powerful energy source.

Fill in the blank with the word or words that best complete each statement.

11. The nuclear reaction in which two nuclei combine is called

    ______________________.

12. Fusion reactions take place naturally on the __________.

13. In a nuclear fusion reaction, hydrogen nuclei combine to form the

    nucleus of a ____________ atom.

Check your answers on page 230.

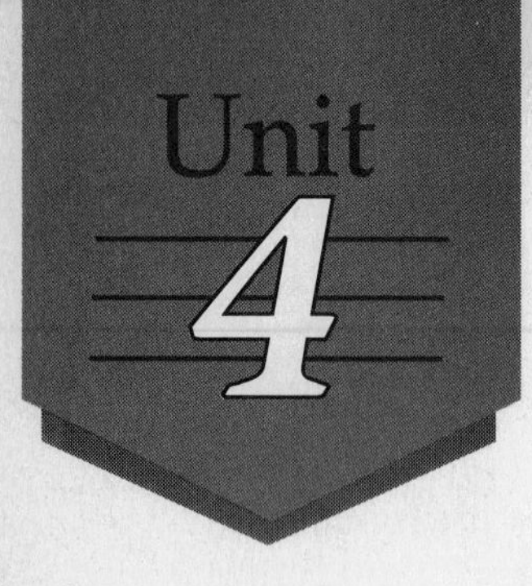

PHYSICS

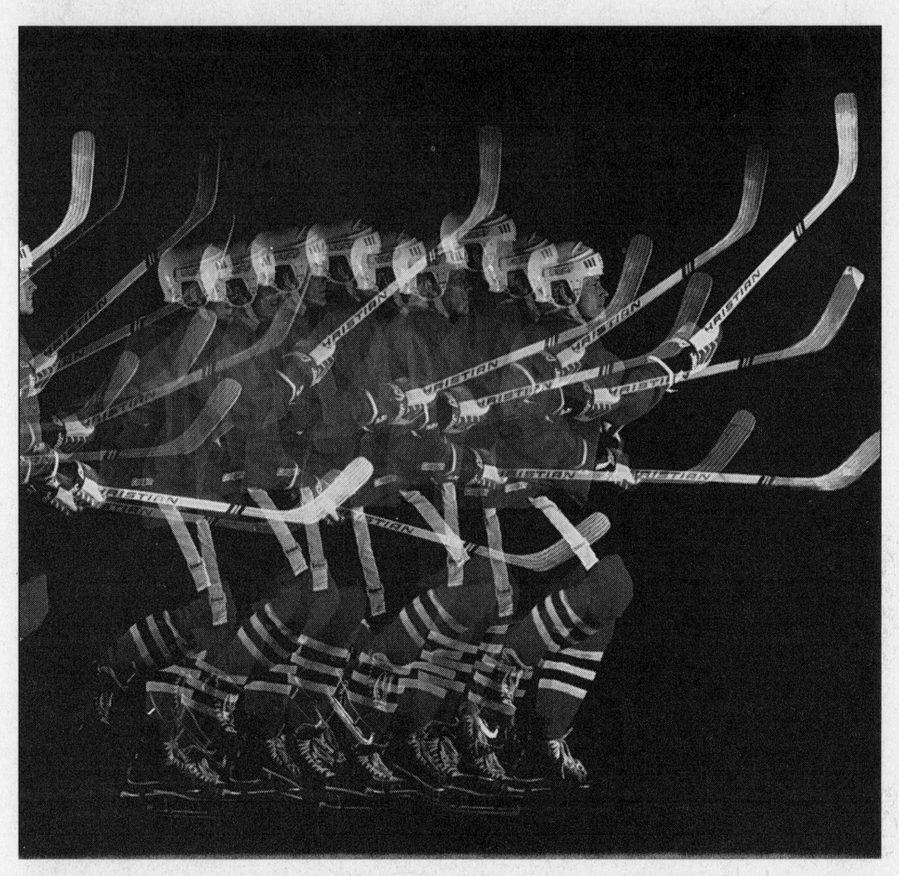

Physics is the branch of science that explains how forces affect motion. The hockey player swings his stick to exert force on the hockey puck.

Physics is the study of energy and how it affects matter. Everything you do requires energy. You use energy when you pick up a pencil and write your name. You use energy when you walk across the street.

Energy comes in many forms. Sound, light, and electricity are forms of energy. Learning about these forms of energy can help you understand why sound and light behave as they do. It can also help you use electricity safely.

Many laws of physics deal with forces and motion. Forces, such as gravity, act on us all the time. Forces make machines work. Understanding forces and machines can help you move or lift heavy objects.

Sometimes forces cause objects to bump into one another, or collide. If you have ever played pool, you may know how to predict where the balls will go after they collide. Understanding what happens in a collision may help make you a safer driver.

Physicists are scientists who study physics. Franklin Chang-Díaz is a physicist with a very special job. He is an astronaut. Chang-Díaz was born in Costa Rica and went to college in the United States. He received his Ph.D. degree in plasma physics. This branch of physics deals with a high-energy state of matter that is found in nuclear reactors. Chang-Díaz has worked on projects studying fusion, a kind of nuclear reaction. He has also worked on designs that use fusion in rocket engines.

Chang-Díaz has been an astronaut since 1981. For the first few years, he worked as part of the ground support team for Spacelab and space shuttle missions. His first space shuttle flight was in 1986. During this mission Chang-Díaz helped place a satellite in orbit. On his second mission, he ran some experiments in the lab on board the shuttle. One of these experiments tested the effects of gravity on plants. Chang-Díaz also carried out a project on growing crystals in space.

Because of his training in science, Chang-Díaz has kept in touch with many scientists. He continues working to bring astronauts and scientists together to solve research problems.

Franklin Chang-Díaz

This unit features articles about several aspects of physics.

- Articles about force and momentum show how forces move machines and how forces and energy are involved in hitting a baseball.
- An article about lasers explains how lasers work and how they are used by doctors.
- An article about electricity discusses how current flows through a circuit and how fuses work.

Machines

Setting the Stage

Walk into a bicycle store, and you'll see an amazing assortment of bikes. There are racing bikes and mountain bikes. There are one-speed bikes and 21-speed bikes. Despite these differences, all bicycles contain the same simple machines.

Past: What you already know

You may already know something about bicycles or machines. Write three things you already know.

1. ______________________________

2. ______________________________

3. ______________________________

Present: What you learn by previewing

Write the headings from the article on pages 173–175 below.

How a Bicycle Works

4. ______________________________

5. ______________________________

6. ______________________________

What do the drawings near the top of page 174 show?

7. ______________________________

Future: Questions to answer

Write three questions you expect this article to answer.

8. ______________________________

9. ______________________________

10. ______________________________

Check your answers on page 230.

How a Bicycle Works

As you read each section, circle the words you don't know. Look up the meanings.

In large cities, businesses use messengers to carry documents across town. These messengers can be seen speeding past clogged traffic. Are they running? No, they're riding bicycles.

A bicycle is a good way to get around. You don't have to be very strong to go at a moderate speed. If you have a bicycle with more than one speed, it can help you tackle hills. By riding a bicycle, you can get your exercise on the way to work.

A bicycle is a good means of transportation.

Force and Work

When you push down on the pedal of a bicycle, you are applying a force to it. A **force** is a push or a pull. There are many kinds of forces. The pull of one magnet on another is a force. The pull Earth has on you and all objects is the force of **gravity**. A force between surfaces that touch each other is **friction**. Friction between the brake pads and a bicycle wheel stops the turning of the wheel.

Work is done when a force causes an object to move. Lifting a bag of groceries is work. Your upward force of lifting overcomes the downward force of gravity. The force you exert is called the **effort**. The force you overcome is the **resistance.**

Simple Machines

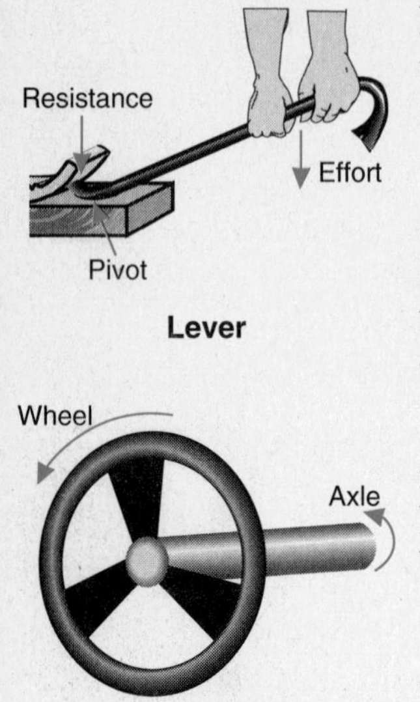

Lever

Wheel and Axle

Sometimes your effort cannot overcome the resistance. Then you need help. If you have ever used a screwdriver to pry open a paint can, you have used a machine. A **simple machine** is a device used to do work. There are several kinds of simple machines. A bicycle contains three of them.

A **lever** is a bar that turns on a pivot. A **pivot** is an object on which another object turns. When you open the paint can, you are using the screwdriver as a lever. The lever takes the effort force from your hand and transfers it to the lid. The lever also multiplies your force. This force overcomes the friction holding the lid on the can.

The number of times a machine multiplies your effort is the **mechanical advantage** of the machine. You may have noticed that a longer screwdriver is a better lever than a short screwdriver. That is because a longer lever gives a greater mechanical advantage.

A **wheel and axle** is made of two objects that turn on the same center. A wheel and axle doesn't always look like a wheel. A crank handle is a wheel and axle. Pencil sharpeners also work this way. You turn the handle through a big circle. This turns the blades inside the sharpener in a small circle. As a result, some wood is cut away, and the pencil gets sharper.

Applying Your Skills and Strategies

Recognizing Cause and Effect. Situations in which one thing makes another thing happen are cause-and-effect relationships. For example, when you turn a crank handle, the blades inside a pencil sharpener turn. Your action is the *cause*. The turning of the blades is the *effect*. An effect can cause something else to happen. When the blades inside the pencil sharpener turn, what is the effect on the pencil?

Gears are wheels with teeth. Unlike a wheel and axle, gears do not turn on the same center. The gears are arranged so that the teeth meet, and each gear turns on its own center. If you turn one gear, you cause any gear it touches to turn.

A Compound Machine

Sometimes it takes more than one machine to get a job done. A **compound machine** is made up of several simple machines. A bicycle is a compound machine. The brake handles are levers and gear shifters. The wheels and pedals are wheel and axle machines. Bicycles also contain gears. The number of speeds a bicycle has equals the number of front gears multiplied by the number of back gears. With many speeds, you can get the exact mechanical advantage you need for any situation.

Check your answer on page 230.

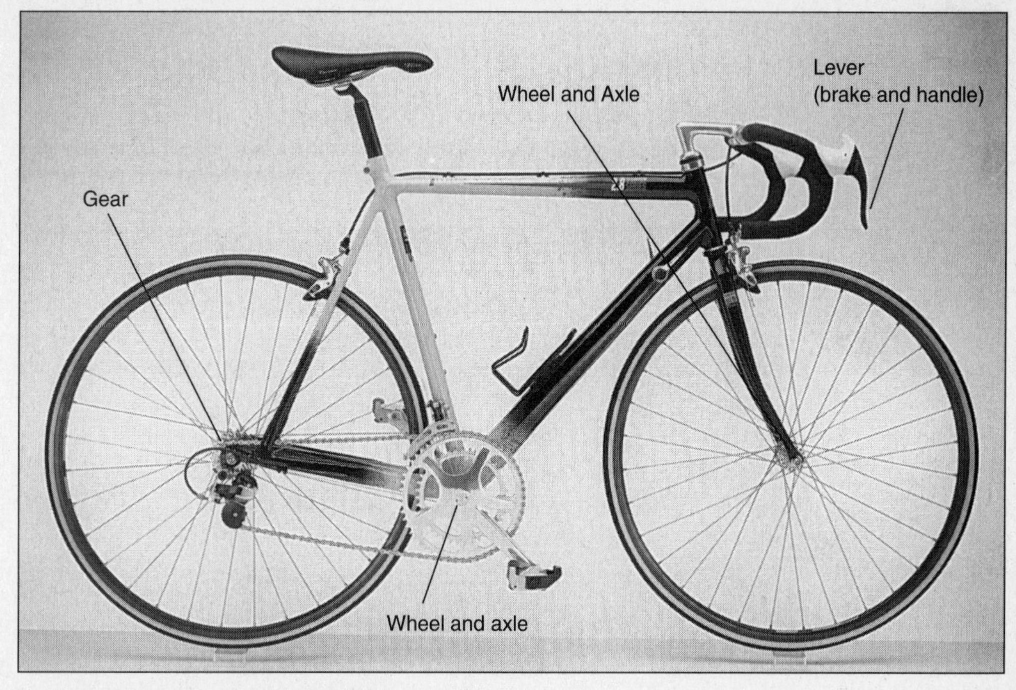

A bicycle is a compound machine.

A machine can multiply your effort. It can also multiply your speed. However, it cannot do both at once. On a level path, you use the higher speeds. The highest speed on a bicycle combines the largest front gear with the smallest back gear. You have to pedal hard but not very quickly. Yet the bicycle moves quickly. The only resistance is friction, which is a small force. The bicycle is not multiplying your force. Instead, it is using your force to multiply your speed.

You use the lower speeds to climb hills. In the lowest speed, you combine the smallest front gear with the largest rear gear. You don't need to pedal hard. However, you do have to pedal quickly, even though the bike moves slowly. The bicycle is multiplying your effort to move you against a large force—gravity.

Applying Your Skills and Strategies

Drawing Conclusions. Conclusions are ideas that are based on facts. They follow logically from the facts. The two paragraphs on this page describe pedaling a bicycle at the lowest and highest speeds. From this you can conclude that if you can pedal quickly, you don't need to use much force. Write another conclusion based on the information on this page.

____

____

____

____

Check your answer on page 230.

Thinking About the Article

Fill in the blank with the word or words that best complete each statement.

1. A __________ is a push or a pull.
2. The force you exert to do work is the __________.
3. The force you overcome when you do work is the

   __________.
4. A device used to do work is called a __________.
5. A __________ machine contains several simple machines.
6. The number of times a machine multiplies your effort is the

   __________ of the machine.
7. The force of __________ is Earth's pull on you and all objects.
8. Wheels with teeth are called __________.

Write your answers in the space provided.

9. Review the questions you wrote on page 172. Did the article answer your questions? If you said *yes*, write the answers. If your questions were not answered, write three things you learned from this article.

____

____

____

____

____

10. What kind of simple machine is the handle that you use to flush a toilet?

____

11. What kind of simple machine is a doorknob?

____

Check your answers on page 230.

Circle the number of the best answer.

12. A machine multiplies
    (1) only speed.
    (2) only force.
    (3) both speed and force at the same time.
    (4) speed at one time and force at another time.
    (5) neither speed or force.

13. When gears turn, they rub on each other. This adds an extra resistance, which is caused by the force of
    (1) friction.
    (2) gravity.
    (3) magnetism.
    (4) effort.
    (5) advantage.

14. If the resistance force is larger than the effort force, no work is done because work is done only
    (1) if an object moves.
    (2) when there is no resistance.
    (3) when there is no effort.
    (4) if there is friction.
    (5) if there is gravity.

Write your answers in the space provided.

15. When some people ride a bicycle up a hill, they stand up as they pedal. How does this help them get up a hill?

____

____

16. Describe an experience you or someone you know has had riding or repairing a bicycle.

____

____

____

Check your answers on pages 230–231.

Momentum

Setting the Stage

The sound of a baseball bat making contact with the ball is unmistakable. This collision sends the ball flying. If the ball takes the right path, there could be a home run. Some players are willing to bend the rules to make this happen.

Past: What you already know

You may already know something about hitting a baseball or other kinds of collisions. Write three things you already know.

1. ______________________________

2. ______________________________

3. ______________________________

Present: What you learn by previewing

Write the headings from the article on pages 179–181 below.

Breaking the Rules

4. ______________________________

5. ______________________________

6. ______________________________

What does the photo on page 179 show?

7. ______________________________

Future: Questions to answer

Write three questions you expect this article to answer.

8. ______________________________

9. ______________________________

10. ______________________________

 Check your answers on page 231.

Breaking the Rules

As you read each section, circle the words you don't know. Look up the meanings.

A major-league pitcher can throw a fastball at more than ninety miles per hour. At this speed, the ball reaches the batter in less than half a second. The batter has only a bit more than one tenth of a second to decide if the pitch looks good. The batter must swing quickly. The swing cannot be early or late by more than a few thousandths of a second. If it is, what might have been a home run becomes a foul ball.

Even good batters get hits only three out of ten times at bat. So batters are always looking for ways to improve their averages. Practice helps. Physical conditioning is also important. However, some batters look for ways that are outside the rules. Most of these ways involve changing the bat.

Collisions

A moving baseball has energy. A catcher can feel this energy as the pitched baseball slams into the mitt. In a **collision**, a moving object strikes another object. The second object may or may not be moving. A catcher's mitt is not moving at the time the ball collides with it. A bat, on the other hand, is moving as the ball collides with it.

These players have just collided at home plate.

All moving objects have momentum. **Momentum** depends on the object's weight and speed. When objects collide, momentum is transferred from one object to the other. Suppose you stand still with your arm extended to the side. If someone throws a baseball into your hand, this collision will push your hand back. The ball transfers some of its momentum to your hand. The effect of the ball on your hand would be greater if the ball had more momentum. This would be true if the ball were heavier or moving faster.

The transfer of momentum is more complicated when both objects are moving. When the moving bat hits the moving ball, they are traveling in opposite directions. The bat is not moving as fast as the ball, but it is much heavier. The bat has more momentum than the ball. When the two collide, the ball moves off in the direction in which the bat was swinging.

The transfer of momentum also depends on how elastic the objects are. Something that is **elastic** can be stretched or compressed and will return to its original shape. In many collisions much of the momentum is lost. If the colliding objects are very elastic, only a little momentum is lost. If you drop a golf ball and a Super Ball, the Super Ball will bounce higher. This is because the Super Ball loses much less momentum.

Applying Your Skills and Strategies

Drawing Conclusions. A conclusion is an idea that follows logically from the information you have. Conclusions must be supported by facts. You have just read that a baseball moves in the direction in which the bat is swinging. You have also read that the bat has more momentum than the ball. You could conclude that colliding objects move in the direction of the object that has the most momentum. Reread the previous paragraph. What can you conclude about how elastic a Super Ball is?

____

____

____

Corking Bats

Some baseball players "cork" their bats. The top of the bat is cut off and the bat is hollowed out. The space is filled with cork, sawdust, or even Super Balls. Then the top of the bat is glued back on. Baseball players feel that this kind of change makes a bat springy. They think the bat becomes more elastic.

Corking bats is against the rules of professional baseball. If a player who hasn't been hitting well suddenly hits a string of home runs, the bat may be taken by the umpires. The bat is x-rayed or cut open. If the bat is corked, the player may be suspended from playing.

Check your answer on page 231.

When Howard Johnson's hitting suddenly improved, he was accused of corking his bat. The bat was x-rayed and found to be solid wood.

Is Cheating Worth It?

Players who cork their bats take a risk. The player who used Super Balls got caught when his bat cracked. The balls bounced out right in front of the umpire! Many players believe it is worth risking a suspension. After retiring, one player admitted to using a corked bat for years. He believed that it added many home runs to his statistics.

Scientists have analyzed what happens when a bat is corked. They have found that the bat gets lighter. In the opinion of some scientists, this weight loss is the reason that the corked bat is better. A batter can swing a lighter bat more quickly. If the batter can swing more quickly, the swing can be started a bit later. This gives the batter a little more time to decide whether or not to swing at the ball.

Applying Your Skills and Strategies

Distinguishing Fact from Opinion. Facts can be proved true. Opinions, on the other hand, are what someone believes, thinks, or feels. They may or may not be true. When reading about science topics, it is important to distinguish fact from opinion. Reread the previous two paragraphs. Write one fact and one opinion from these paragraphs.

____

____

____

____

There may be one more advantage to the corked bat. This one is in the player's mind. If the batter thinks that the bat is an advantage, it may improve the batter's confidence. The batter steps up to the plate believing that a home run is about to happen.

Check your answer on page 231.

Thinking About the Article

Fill in the blank with the word or words that best complete each statement.

1. In a ________________, a moving object strikes another object.
2. An object's ________________ depends on its weight and speed.
3. An object that is ______________ returns to its original shape after being stretched or compressed.

Write your answers in the space provided.

4. Review the questions you wrote on page 178. Did the article answer your questions? If you said *yes*, write the answers. If your questions were not answered, write three things you learned from this article.

____

____

____

____

____

5. How does a player cork a bat?

____

____

____

____

Circle the number of the best answer.

6. How can you increase an object's momentum?
   (1) Increase its weight.
   (2) Decrease its size.
   (3) Increase its speed.
   (4) options (1) and (2)
   (5) options (1) and (3)

 Check your answers on page 231.

7. All collisions are alike in that one object is moving
   (1) and it is moving at the same speed as the other object.
   (2) and the other object is not moving.
   (3) quickly and the other is moving slowly.
   (4) and the other object may or may not be moving.
   (5) and then both objects stop.

8. Which of the following is a fact?
   (1) Batters believe that a corked bat helps them hit the ball better.
   (2) Some people think that a corked bat helps a player's confidence.
   (3) Scientists have found that corking a bat makes it lighter than a solid bat.
   (4) One player feels that a corked bat gave him many home runs.
   (5) Scientists think that the loss of weight is what makes a corked bat better than a solid bat.

Write your answers in the space provided.

9. If a compact car and a limousine are moving at the same speed along a highway, which car has more momentum? Explain your answer.

___

___

___

10. Think about what a car looks like after it has been in a collision. How can you tell that the body is not very elastic?

___

___

11. Many movies and TV programs have car chase scenes that end in collisions. Describe such a collision and any transfer of momentum that you could see.

___

___

___

___

Check your answers on page 231.

Electricity

Setting the Stage

It is hard to imagine living without electricity. It powers lamps, fans, heaters, air conditioners, refrigerators, radios, and TVs. If you have ever blown a fuse, you know that there is a limit to how many of these things you can use at one time.

Past: What you already know

You may already know something about electricity or fuses. Write three things you already know.

1. ______________________

2. ______________________

3. ______________________

Present: What you learn by previewing

Write the headings from the article on pages 185–187 below.

Using Electricity Safely

4. ______________________

5. ______________________

6. ______________________

What does the photo on page 185 show?

7. ______________________

Future: Questions to answer

Write three questions you expect this article to answer.

8. ______________________

9. ______________________

10. ______________________

Check your answers on page 231.

Using Electricity Safely

As you read each section, circle the words you don't know. Look up the meanings.

Electricity seems like magic. Plug in a tape player and you hear music. Plug in a lamp and you have light. However, many people use electricity in unsafe ways. Most of the time this is caused by trying to use too much electricity at once.

What Is Electricity?

Electricity is a kind of energy. It flows through power lines on poles and through the wires in your home. Electricity that flows through a wire or another object is called an **electric current**. When a current passes through certain objects, the electrical energy is changed into another form of energy. In a lamp, the change is to light energy. In a tape player, the change is to sound energy.

Materials that carry current are called **conductors.** Most metals are good conductors. Copper, which is used in home wiring, is a very good conductor. Materials that do not carry current are called **insulators**. Glass, wood, and plastic are insulators.

These wires are part of the circuit that carries electric current to the people who use it.

Electric current is produced at a power plant. Wires carry the current to homes, stores, and factories. Wires form a **circuit**, or a path through which electricity can travel. When you turn on a lamp, you add its wires to the circuit. Electricity flows through the circuit, lighting the lamp.

Current will flow only if the circuit is complete, or closed. If a connection in the circuit is opened, the path for the current is broken, and the current cannot flow. This is how a switch is used to turn a lamp or appliance on and off.

Volts and Amps

Electricity is measured in several ways. **Voltage** is a measure of how much energy is available to the whole circuit. Voltage is similar to water pressure. Every pipe in a water system has the same amount of pressure on it. Pressure is what pushes water through the pipes. Voltage is what pushes current through a circuit. Voltage can be thought of as electrical pressure.

In the United States, household circuits are usually 110 or 120 volts. Electrical devices are made to work on this voltage. In Europe, voltages are higher. What happens to a water pipe if the pressure suddenly increases? It may burst. What if you use a device designed for 110 volts in a circuit with a higher voltage? It doesn't burst, but it usually overheats. If some of the wires melt, the device may be ruined. Many American tourists have destroyed their hair dryers while on a European vacation!

Applying Your Skills and Strategies

Making Inferences. An *inference* is a fact or an idea that follows logically from what has been said. It is important to make inferences as you read. Reread the previous paragraph. Based on the facts in that paragraph, you can infer that electricity produces heat. What can you infer about appliances made in Europe?

____

____

____

The amount of current that flows is called **amperage**. Current is measured in amperes, or amps. Each appliance uses a certain number of amps. The more appliances you add to a circuit, the more amps of current there will be flowing through the circuit. Again, think of water in pipes. Turning on a faucet so that it trickles uses a small amount of water. And, using a few appliances uses only a few amps of current.

What happens if you plug in more appliances? If you turn on the faucet all the way, you use more water. Opening the faucet more causes you to use more water. Turning on more appliances causes you to use more amps of current.

Too Much of a Good Thing?

Can you use too many amps of current? If you open all the faucets, nothing bad happens to the pipes. This is where the comparison to water ends. If you plug in and turn on too many devices, something very bad will happen to your circuits. They will become overloaded.

Check your answer on page 231.

When current flows through wires, it causes them to become warmer. A toaster is designed to make use of this effect. But the wires in the walls of your home are not designed for this heat. If you use too many appliances, too much current may flow. The circuit may heat up. If the wires get too hot, they could start a fire.

Fuses keep overloads from happening. Fuses are rated in amps. The number of amps marked on the fuse should match the number of amps that the circuit is designed to carry. If you use too much current, the heat melts a thin strip of metal in the fuse. This opens the circuit. Current stops flowing, and the circuit cools off. A circuit breaker is like a fuse. But it has a heat-activated switch in place of the thin strip of metal.

Fuses come in a variety of shapes and sizes. When the metal in a fuse melts, the circuit is opened and no current flows.

Applying Your Skills and Strategies

Recognizing Cause and Effect. Situations in which one thing makes another thing happen are *cause-and-effect* relationships. For example, if too much current flows, the circuit heats up. The current is the cause and the heat is the effect. What is the effect of the heat on the fuse?

____

____

People get annoyed when fuses blow. First, they have to unplug some of the appliances. Then they have to go to the fuse box, take out the old fuse, and put in a new one. The next time they plug in all the appliances, the fuse blows again. Some people avoid this nuisance by using fuses rated for more amps. If you put a 20-amp fuse in a 15-amp circuit, you can plug in more appliances without blowing the fuse. But changing the fuse doesn't change the wires. The higher-rated fuse may not blow until after the circuits are dangerously hot. This is why you should never use a fuse that has a higher rating than the circuit.

Check your answer on page 231.

Thinking About the Article

Fill in the blank with the word or words that best complete each statement.

1. Electricity that flows through a wire or another object is called an

   ______________________________.

2. The path along which an electric current travels is a ______________.

3. ______________ is a measure of how much energy is available to a circuit.

4. The amount of current that flows is the ______________.

Write your answers in the space provided.

5. Review the questions you wrote on page 184. Did the article answer your questions? If you said *yes*, write the answers. If your questions were not answered, write three things you learned from reading this article.

____

____

____

____

____

6. Name two safety devices used in electric circuits.

____

____

Circle the number of the best answer.

7. Which of the following is an insulator?

   (1) copper

   (2) aluminum

   (3) plastic

   (4) steel

   (5) silver

Check your answers on page 232.

8. Which of the following is a form of energy?
   (1) electricity
   (2) sound
   (3) light
   (4) heat
   (5) all of the above

9. If you replaced a 15-amp fuse with a 10-amp fuse, what would be the effect on the circuit?
   (1) You could use more appliances safely.
   (2) It would take fewer appliances to blow the fuse.
   (3) It would cause a fire.
   (4) You would have to use foreign appliances.
   (5) There would be no effect.

Write your answers in the space provided.

10. Before working on the wiring of your home, an electrician removes the fuses or opens the circuit breakers. Why is this important?

___

___

___

11. Where are the fuses or circuit breakers located in your home?

___

___

12. Describe a problem you or someone you know has had with an overloaded circuit or an electric appliance. How was the problem solved?

___

___

___

___

___

Check your answers on page 232.

Light

Setting the Stage

What do you think of when you hear the word *laser*? Perhaps you think of special effects at a rock concert. Maybe you think of science-fiction characters with laser weapons. You may not think of medical tools. Yet lasers and the light that they produce have many uses in medicine.

Past: What you already know

You may already know something about light or lasers. Write three things you already know.

1. ____________________

2. ____________________

3. ____________________

Present: What you learn by previewing

Write the headings from the article on pages 191–193 below.

Amazing Lasers

4. ____________________

5. ____________________

6. ____________________

What does the diagram on page 192 show?

7. ____________________

Future: Questions to answer

Write three questions you expect this article to answer.

8. ____________________

9. ____________________

10. ____________________

 Check your answers on page 232.

Amazing Lasers

As you read each section, circle the words you don't know. Look up the meanings.

In science-fiction movies, the hero "zaps" the villain with a laser gun. However, lasers are not really used as guns. Lasers are used as tools. A **laser** produces a narrow, strong beam of light. In science, laser light is used for making measurements. In industry, laser light is used to cut things. Laser light can cut pieces of steel. It can also cut fancy patterns out of a piece of paper.

Laser light is being used more and more by doctors. Laser light can remove birthmarks and tattoos. It can break up gallstones and tumors. Laser light can even repair a damaged eye. Different kinds of lasers are used for each purpose. However, they all work using the same basic design.

Getting in Step

Light is a form of energy that travels in waves. Other kinds of waves include sound waves, radio waves, and x-rays. A wave has a regular pattern. It is usually drawn as a curved line that goes up and down. If you drop a stone into a pond, ripples of water move out in a circle. Each ripple is a wave. The energy of the wave makes the water move up and down. The **frequency** is the number of waves that pass a point in a certain amount of time. The **wavelength** is the distance from the top of one wave to the top of the next wave.

What we see as ordinary white light is made up of all the colors of a rainbow. Each color of light has its own frequency and wavelength. Imagine a crowd of people moving forward at the same speed. The people are not trying to walk in step. This is what ordinary light is like. Now imagine the crowd all marching in step. This is what laser light is like. All the waves have the same frequency and wavelength.

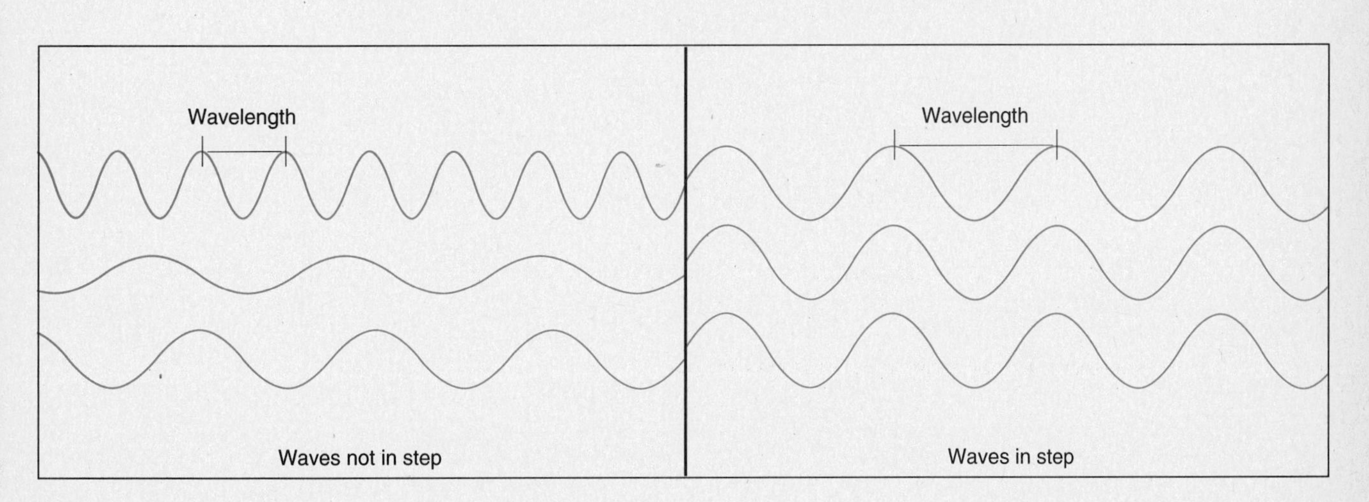

How a Laser Works

A laser contains a material that will give off light waves that are all in step. A mirror at one end of the tube reflects light completely. A mirror at the other end reflects only some light. A burst of electricity or light is set off inside the tube. This energy is absorbed by the atoms in the tube. The atoms then give off the energy in the form of light. All the light given off has the same wavelength and frequency. The light is reflected back and forth between the mirrors. The light gets stronger and stronger as this happens. When the light becomes strong enough, it passes through the partially reflecting mirror. This light is the laser beam. What makes one kind of laser different from another is the material in the tube. Each kind of atom gives off a different wavelength of light.

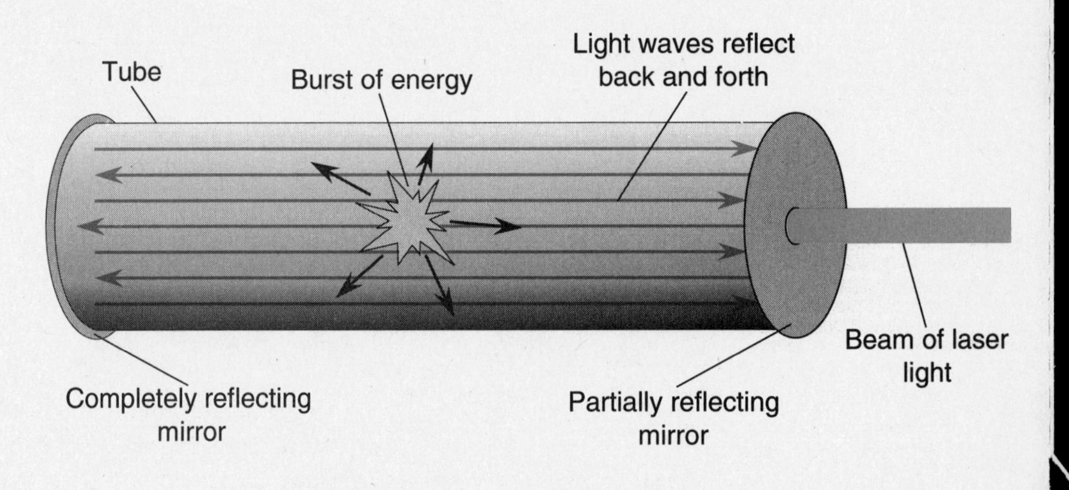

How a Laser Produces Light

Applying Your Skills and Strategies

Reading a Diagram. A diagram can help you see how something looks or works. This helps you follow what the author is describing. The title tells you the content of the diagram. Circle the title of the diagram on this page. Labels help you identify certain parts of the diagram. Find the labels that show the two mirrors in the laser. Underline these labels. Through which mirror does the light leave the laser?

____

____

____

The light from a laser is very strong. One reason is that the waves are all in step. Another reason is that the light beam does not spread out very much. Since laser light is so strong, it can cut through metal. Laser light can damage your eyes if you look directly at it. That is why the laser lights at rock concerts are aimed upward and not at the audience.

Check your answer on page 232.

Lasers in Medicine

When laser light is directed at an object, the object absorbs the energy. The energy makes the object warmer. The heat made by laser light can be used in many ways.

A broken blood vessel inside the eye can damage a person's vision. Because the laser beam is so narrow, it can be aimed inside the eye at the exact spot where the blood vessel is broken. The heat caused by the laser beam seals off the broken blood vessel. Yet it does not harm the rest of the eye. Different substances absorb different wavelengths. The laser light used on the eye is a wavelength that the blood vessel absorbs. This wavelength is not absorbed by other parts of the eye.

Applying Your Skills and Strategies

Summarizing Information. Summarizing means briefly telling the main ideas. When you summarize, you leave out many details. You tell only the important things. Reread the previous paragraph. Underline the important points. Now write a summary of the paragraph in one sentence.

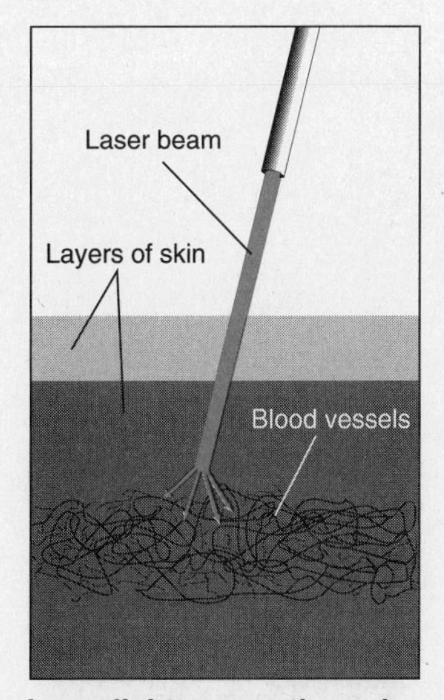

Laser light passes through the skin and is absorbed by small blood vessels.

About three in every thousand children have birthmarks call port-wine stains. These birthmarks are large patches of very red skin. They are caused by abnormal blood vessels in the skin. Doctors have tried for many years to use lasers on port-wine stains. The laser light destroyed the poorly formed blood vessels. But the laser beam also caused scars in the normal layers of skin above the blood vessels.

Now there are lasers that produce light in very short pulses. These pulses last less than one thousandth of a second. In that short time, they destroy the blood vessels without harming the skin. This treatment has been tried on many children. Their port-wine stains were removed in about six treatments spread out over one year.

The pulsed laser may soon become a dental tool. Laser light can remove decay without hurting the rest of the tooth. But light from an ordinary laser heats the gums too much. For this reason, dentists have not been able to use lasers. Now some dentists are experimenting with the new pulsed lasers. The short burst of light should destroy the decay without damaging other tissues. Someday you may go to your dentist to have a cavity "zapped"!

Check your answer on page 232.

Thinking About the Article

Fill in the blank with the word or words that best complete each statement.

1. __________ is a form of energy that travels in waves.

2. The __________________ is the number of waves that pass a point in a certain amount of time.

3. The ____________________ is the distance from the top of one wave to the top of the next wave.

Write your answers in the space provided.

4. Review the questions you wrote on page 190. Did the article answer your questions? If you said *yes*, write the answers. If your questions were not answered, write three things you learned from reading this article.

____

____

____

____

____

5. How is laser light different from ordinary white light?

____

____

____

Circle the number of the best answer.

6. Which of the following is not part of a laser?
   (1) a partially reflecting mirror
   (2) a completely reflecting mirror
   (3) a light bulb
   (4) atoms that give off light
   (5) a source of light or electricity

Check your answers on page 232.

7. When laser light is absorbed by body tissues, they
   (1) get colder.
   (2) get hotter.
   (3) give off light.
   (4) give off electricity.
   (5) grow larger.

8. Why can't the same laser be used to cut metal and repair an eye?
   (1) The metal and the eye absorb different wavelengths of light.
   (2) The metal cannot be heated as much as the eye can.
   (3) It takes a larger laser to repair an eye.
   (4) Only the light used to repair an eye should be all in step.
   (5) Only the light used to cut metal should be all in step.

Write your answers in the space provided.

9. How are the effects of the pulsed laser different from the effects of other lasers?

____

____

____

10. Give two reasons why laser light is so useful in repairing damage inside the eye.

____

____

____

11. Suppose you had a cavity in a tooth and the dentist suggested removing it with a laser instead of a drill. Would you try the laser? Explain your answer.

____

____

____

____

Check your answers on pages 232 - 233.

Unit 4 Review:
Physics

The Inclined Plane

A **simple machine** is a device to do work. An inclined plane is a simple machine. An **inclined plane** is a long, sloping surface that helps raise an object that cannot be lifted. A ramp is an example of an inclined plane. Imagine trying to get a person in a wheelchair onto a porch that is one foot above the ground. Lifting the person and the wheelchair would be difficult. You may not be able to exert a large enough force to do this. If you had a ramp, you could easily get the person and the wheelchair onto the porch.

The amount of work done depends on the force needed and the distance moved. You would need less force to move the person up the ramp. However, you would have to move the person a longer distance, four feet instead of one. It takes about the same amount of work to push the person up the ramp as it does to lift the person.

Friction is a force between surfaces that rub against each other. Friction makes it harder to use an inclined plane. It is easy to use an inclined plane to move the person in a wheelchair because the wheels have little friction. But if you had to push a box up the same ramp, you would need a large force just to overcome the friction.

Fill in the blank with the word or words that best complete each statement.

1. A ______________________________ is a device to do work.
2. An ______________________________ is a long, sloping surface that helps raise an object that cannot be lifted.
3. ____________________ is a force between surfaces that rub against each other.

Circle the number of the best answer.

4. In which situation would you need to exert the least force?
   (1) lifting a 75-pound box to a height of one foot
   (2) lifting a 120-pound box to a height of one foot
   (3) pushing a 75-pound box up a ramp three feet long
   (4) pushing a wheeled 75-pound cart up a ramp three feet long
   (5) lifting a wheeled 75-pound cart up to a height of one foot

Go on to the next page.

Momentum

If an object is moving, it has momentum. The object's **momentum** depends on its weight and speed. The faster an object moves, the more momentum it has. If an object is not moving, it has no momentum.

Momentum is transferred between objects when they collide. Pool players make use of the transfer of momentum. A good player knows how to set up collisions between the balls on the table. The player uses a long stick, the cue, to hit one of the balls. The ball that is hit is called the cue ball. In the collision between the cue and the cue ball, momentum is transferred to the ball. The cue ball rolls along the table until it strikes one of the numbered balls. The momentum of the cue ball is transferred to the numbered ball. The cue ball stops moving and the numbered ball moves. If the player has aimed correctly, the numbered ball will go into a pocket.

A good pool player knows how to set up collisions between more than two balls. If the cue ball strikes two numbered balls at the same time, the two balls move away from the cue ball. Each of the numbered balls gets some of the momentum of the cue ball. Neither of the numbered balls moves as fast as the cue ball was moving.

Players also make bank shots, in which a ball collides with the side of the table. In this collision, the ball keeps its momentum and bounces off the side of the table in a new direction.

Circle the number of the best answer.

5. What factors determine an object's momentum?
   (1) speed and direction of motion
   (2) weight and direction of motion
   (3) weight and speed
   (4) weight and how far the object has moved
   (5) speed and how far the object has moved

6. What happens to a rock as it rolls down a hill?
   (1) Its momentum increases because the speed increases.
   (2) Its momentum decreases because the speed increases.
   (3) Its momentum increases because the rock gets heavier.
   (4) Its momentum decreases because the rock gets heavier.
   (5) Its momentum does not change.

Go on to the next page.

Sound Waves

Sound is caused by vibrations. If you pluck a stretched rubber band, you can see it vibrate, or move back and forth. The rubber band makes the air near it vibrate. A vibrating object, such as a rubber band, pushes molecules in the air back and forth. These vibrations move through the air like ripples in a pond. Each ripple is a wave.

If you look at water waves, you can see the distance from the top of one wave to the top of the next. This distance is the **wavelength**. The wavelength of a sound wave can vary. Each wavelength makes a sound of a different pitch. Short wavelengths are high sounds, like the sound of a flute. Long wavelengths are low sounds, like the sound of a tuba.

The height of a wave is the **amplitude**. The amplitude of a sound wave is the loudness of the sound. The higher the amplitude, the louder the sound.

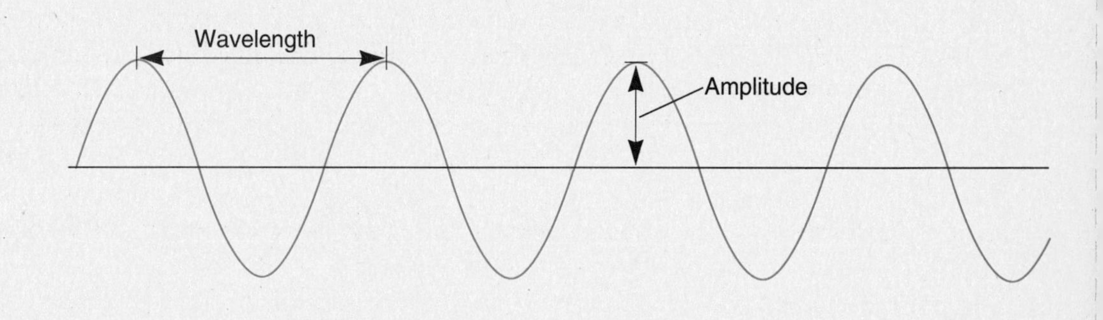

Fill in the blank with the word or words that best complete each statement.

7. __________ is caused by vibrations.

8. The height of a wave is the ________________.

Circle the number of the best answer.

9. A loud, high-pitched sound has a wave with a
   (1) high amplitude and long wavelength.
   (2) high amplitude and short wavelength.
   (3) low amplitude and long wavelength.
   (4) low amplitude and short wavelength.
   (5) low amplitude and medium wavelength.

 Go on to the next page.

Electromagnetic Fields

Any device that uses electricity has an energy field around it. The field caused by an electric current is called an **electromagnetic field**, or **EM field**. The EM field has properties of electricity and magnetism. For example, the EM field of a TV can turn the needle of a magnetic compass, just as a magnet can. An EM field is produced only while the electric current is flowing. When the TV is turned off, there is no EM field.

An EM field can ruin a video- or audiotape. The tape contains metal atoms which are lined up in a pattern. This pattern is read when you play back the tape. Because the EM field can act like a magnet, it can move the metal atoms in the tape. The EM field rearranges the pattern. When you play back the tape, the picture or sound will be scrambled. Computer disks and video games work in the same way as the tapes. This is why tapes, disks, and game cartridges should be kept away from EM fields. Records and compact discs do not have a magnetic pattern, so they are not damaged by EM fields.

The metal detector that people walk through at an airport has an EM field. When you walk through the EM field, metal objects you are carrying affect the field. If the field detects enough metal, an alarm sounds. After removing keys, pocket change, or heavy jewelry, people can walk through the gate without affecting the EM field. The alarm does not ring.

Fill in the blank with the word or words that best complete each statement.

10. An ______________________________ is caused by flowing electricity.

11. An EM field can affect metal because the field acts like a

    ____________.

Circle the number of the best answer.

12. Which of the following can be safely carried through an airport metal detector?

    (1) an audiotape

    (2) a videotape

    (3) a computer disk

    (4) a video game cartridge

    (5) a compact disc

Check your answers on page 233.

POSTTEST

The Scientific Method

Have you ever solved a problem by trial and error, by observation, or by making an informed choice? If so, you have used steps of the scientific method. The scientific method is an organized way of solving problems. Suppose that you have an old table. You suspect that under the scratched, dull finish there is a beautiful wood. How can you remove the old finish?

You have already defined your problem, which is the first step of the scientific method. Now you need to make some observations. The best way to start is to do some reading. Perhaps a friend has books you can borrow. You can also try the library. Another way to collect information is to ask people who have experience with refinishing wood. Again, a friend might be able to help. You could also go to a hardware store, and ask some questions. You could read the labels of several products.

By now, you should have narrowed down your choices to two or three products. One way to decide which one to use is to test them. This is an experiment. Try each product on a small area that doesn't show.

After you try the products, compare the results. This is collecting data. After you look at your data, you should be able to choose the product you want to use. You are drawing a conclusion.

Fill in the blank with the word or words that best complete each statement.

1. Suppose you want to have a surprise party for a friend, but you don't know what to serve. If you call some other friends for suggestions, you are making ____________________.
2. When you realized that you didn't know what to serve, you stated your ____________.
3. One friend suggests angel food cake. You've never made this before, so you try out a few recipes. This is an ______________.
4. You invite some friends to taste your sample cakes. Their reactions are your ________.
5. Based on your taste tests, you choose a recipe. This is your ______________.

Go on to the next page.

How Viruses Reproduce

A virus is a tiny particle made up of genetic material with a protein coating. Viruses are not cells, and they do not reproduce the way cells do. A virus does not contain the raw materials needed to make more viruses. Instead, it gets these materials from a host cell. Each kind of virus needs a particular host.

To reproduce, a virus takes over its host cell. The virus attaches to its host and injects its genetic material into the cell. The genetic material from the virus takes control. Soon the cell becomes a virus factory. When the cell is full of viruses, it breaks open. The new viruses are released. Each one is able to take over another host cell.

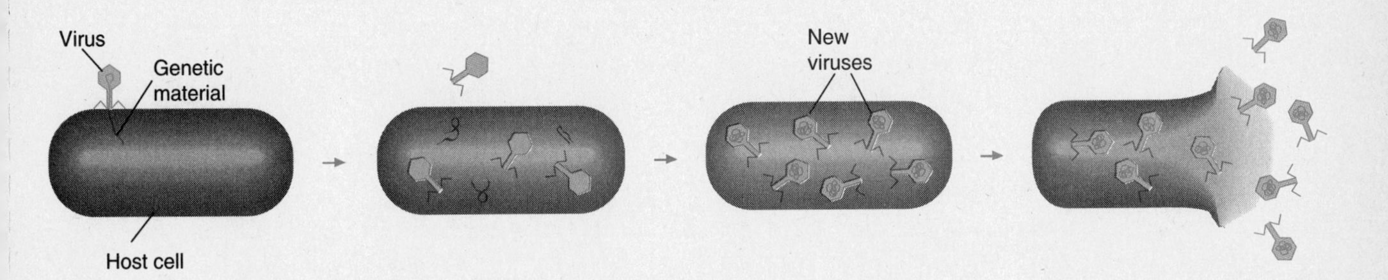

Circle the number of the best answer.

6. A virus needs a host cell to reproduce because the
   (1) virus has the raw materials for making viruses.
   (2) host cell has the raw materials for making viruses.
   (3) virus has no genetic material.
   (4) host cell has the genetic material for making viruses.
   (5) genetic material of the virus and the host are the same.

7. After the new viruses are released, the host cell
   (1) becomes a virus.
   (2) makes several more generations of viruses.
   (3) returns to its original form.
   (4) dies.
   (5) reproduces to make more host cells.

Go on to the next page.

The Predator and the Prey

Snowshoe hares live in the forests of northern Canada. These animals get their name from their large feet. Like snowshoes, large feet spread out the animal's weight. As a result, the hare can walk on snow without sinking into it. Snowshoe hares are caught and eaten by lynxes. A lynx is a large cat. Like the hare, the lynx has large feet that help it walk on top of deep snow.

The lynx is a predator. It kills and eats its prey, the hare. The populations of these two animals are closely linked. The populations grow and shrink in a cycle.

In years when there are many hares, hunting is easy for the lynxes. There is plenty to eat. As a result, the lynx population increases.

In a year or two, there are many lynxes feeding on the hares. Many hares are caught and eaten, and the hare population goes down sharply. After the hare population goes down, there is less food for the lynxes. Hunting becomes harder, and fewer lynxes survive. The lynx population goes down.

When there are fewer lynxes, fewer hares are eaten. Over the next few years, the hare population grows. Soon there are so many hares that hunting becomes easy for the lynxes, and the cycle repeats.

Fill in the blank with the word or words that best complete each statement.

8. An animal that hunts and kills its food is a ________________.

9. An animal that is hunted by another animal is the ________.

Circle the number of the best answer.

10. The snowshoe hare is also eaten by wolves. If a disease suddenly killed off many wolves, the first change you would see is that
    (1) there would be more hares.
    (2) there would be fewer hares.
    (3) there would be fewer lynxes.
    (4) lynxes would catch the disease.
    (5) hares would catch the disease.

Go on to the next page.

Incomplete Metamorphosis

People love to see a butterfly visit a garden and drink nectar from the flowers. Everyone likes butterflies. They're pretty, and they don't chew up the leaves the way caterpillars do. Yet the caterpillar that you hate to see munching on leaves will become the pretty butterfly. This change is called complete metamorphosis because these two states in the life cycle are so different.

Many insects do not go through such a dramatic change. The young insect, called a nymph, looks much like the adult. This slight change is called incomplete metamorphosis. The grasshopper goes through this kind of change. So does the praying mantis. There are two main differences between the nymph and the adult. Only the adult has wings. Only the adult is able to reproduce.

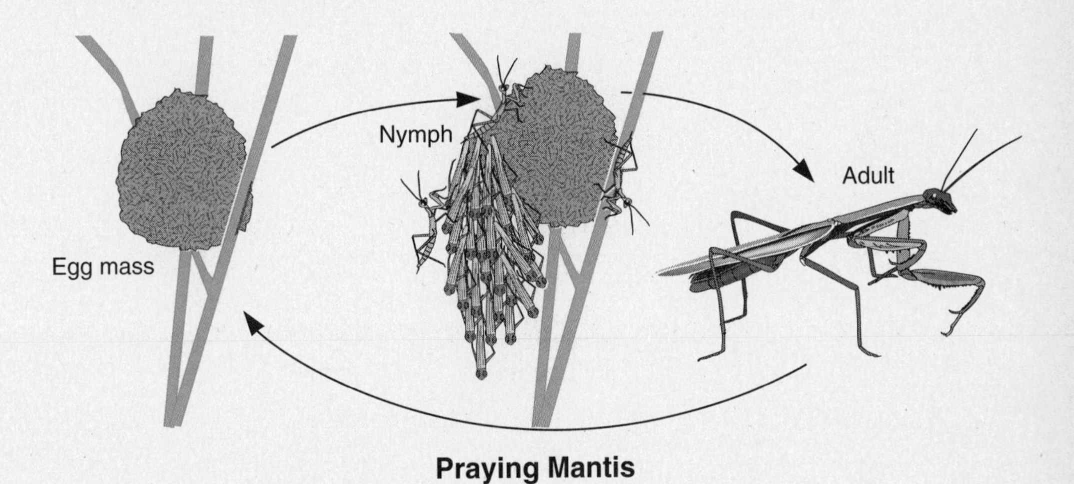

Praying Mantis

Fill in the blank with the word or words that best complete the statement.

11. The change from a nymph to an adult grasshopper is called

_______________________________________.

Circle the number of the best answer.

12. How does a praying mantis nymph differ from the adult?

(1) The adult has wings, and the nymph doesn't.

(2) The adult can eat, and the nymph can't.

(3) The adult can reproduce, and the nymph can't.

(4) (1) and (2)

(5) (1) and (3)

Go on to the next page.

Respiration

Most living things need oxygen to survive. In the process of respiration, cells use oxygen to break down food to give the body energy. Carbon dioxide is given off as a waste product in this process. Respiration goes on in cells all the time. As a result, the body has a constant source of energy. Plants, animals, and some bacteria carry on respiration.

Plants also carry on photosynthesis. This process is the reverse of respiration. In photosynthesis, energy from sunlight is used to change carbon dioxide and water into sugar. The sugar is food for the plant. Oxygen is also produced. Photosynthesis goes on only when the plant receives light.

Plants produce more oxygen than they use. The extra oxygen is given off into the air. Photosynthesis is the source of all the oxygen used by living things.

Circle the number of the best answer.

13. How are photosynthesis and respiration related?

    (1) They are the same.

    (2) They both need oxygen.

    (3) They both need sunlight.

    (4) They both take place in animals.

    (5) Photosynthesis is the reverse of respiration.

14. Animals eat plants to get

    (1) carbon dioxide.

    (2) oxygen.

    (3) food.

    (4) sunlight.

    (5) all of the above.

15. In the process of respiration, plants use

    (1) carbon dioxide.

    (2) water.

    (3) sunlight.

    (4) oxygen.

    (5) all of the above.

Go on to the next page.

Metallic Ores

The steel in the fender of your car came from iron ore. Likewise, the coins in your pocket came from ores of silver and copper. Metallic ores are rocks containing metals that can be mined for profit.

In an ore, the desired substance is chemically combined with another substance. In most metallic ores, the metal is combined with oxygen. Such ores are called oxides. The process of separating the metal from the oxygen and other substances is called refining.

Great heat is needed for refining ores. A substance to combine with the oxygen is also needed. In refining iron ore, coke takes care of both needs. Coke is a form of carbon. The ore and the coke are loaded into a furnace, and the coke is burned. As it burns, it combines with the oxygen from the ore. Carbon dioxide is given off. The metal that is left melts and is collected.

Ores are a resource, a material that people need from Earth. There is only so much of an ore on Earth. When it is used up, there will be no more. This is why metals are being recycled. Scrap iron and steel can be melted down and made into new products. This takes less energy than refining. So recycling helps save two resources—ores and oil.

Fill in the blank with the word or words that best complete each statement.

16. ______________________ are rocks containing metals that can be mined for profit.

17. An ore in which metal is combined with oxygen is an ______________.

18. A material that people need from Earth is a __________________.

Circle the number of the best answer.

19. One of the uses of coke in the refining process is
    (1) to add oxygen to the metal from the ore.
    (2) to add carbon to the metal from the ore.
    (3) to combine with oxygen from the ore.
    (4) to combine with both the metal and the oxygen from the ore.
    (5) none of the above.

Go on to the next page.

Eclipses

On July 11, 1991, an eclipse of the sun took place. Many people traveled to Hawaii and Mexico just to see it. A solar eclipse happens when the moon passes between Earth and the sun. Imagine that you are sitting near a road on a sunny day and a truck goes by. The shadow of the truck passes over you, and you cannot see the sun until the truck passes. This is what happens in a solar eclipse. The moon casts a shadow on Earth.

When you are in the truck's shadow, you can see that areas around you are bright. The truck doesn't block all the light from the sun. It blocks just a small part of the light. This is because the sun is huge, and the truck is so much smaller. In the same way, the moon's shadow falls on only a part of Earth. So the eclipse is seen from only a small part of Earth's surface. Over Hawaii, the sun was completely blocked. This is a total eclipse. Other places had a partial eclipse. Over New York City, only a tiny percent of the sun was hidden by the moon.

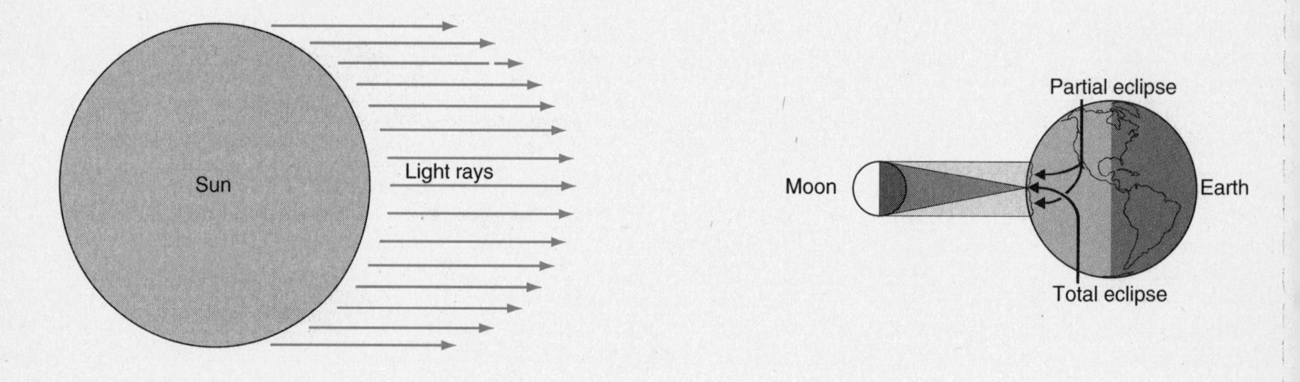

Fill in the blank with the word or words that best complete each statement.

20. A ______________________ happens when the moon passes between the sun and Earth.

21. In places where only part of the sun is blocked, you see a

    ______________________ eclipse.

22. The moon's shadow is small because the __________ is smaller than

    the __________.

Go on to the next page.

Oxidation

In the process of oxidation, a substance reacts with oxygen. The new compound that forms is called an oxide. Many metals oxidize when they are in contact with oxygen in the air.

You have probably seen some metal oxides. For example, iron combines with oxygen, forming iron oxide, or rust. Wrought-iron fences rust. So do the steel bodies of cars. Rust makes the metal look brown and scaly. Rust also weakens the metal. When flakes of rust fall off the metal, they leave more metal exposed to the air. This metal rusts and flakes off. Finally, the metal may rust through, leaving a hole.

Copper also oxidizes. As the red-brown copper oxidizes, it turns green. The green color is a layer of copper oxide. This layer does not flake off as rust does. Instead, the copper oxide stays on the metal and stops any further oxidation.

Aluminum is a metal that does not oxidize easily. That is one reason why aluminum is so useful for making garbage cans or other things that will stay outside in all weather.

Fill in the blank with the word or words that best complete the statement.

23. In the process of ___________________, a substance reacts with oxygen.

Circle the number of the best answer.

24. When silver is tarnished, it has oxidized. Why does covering silver objects in plastic wrap stop this process?

    (1) The plastic wrap keeps the metal warm.

    (2) The plastic wrap keeps the metal dry.

    (3) The plastic wrap keeps the metal away from oxygen.

    (4) The plastic wrap keeps the metal away from rust.

    (5) The plastic wrap keeps the metal away from copper.

25. Which of the following has not been oxidized?

    (1) iron oxide

    (2) copper oxide

    (3) silver oxide

    (4) aluminum oxide

    (5) silver chloride

Go on to the next page.

Acids and Bases

Two large groups of compounds are acids and bases. These compounds can be identified with a substance called an indicator. Litmus paper is an indicator. An acid will turn litmus paper red. A base will turn litmus paper blue. A scale, called the pH scale, tells how strong an acid or a base is. The scale runs from 0 to 14. The strongest acid has a pH of 0. The strongest base has a pH of 14. A substance that is neither an acid nor a base is said to be neutral. Water is neutral. Its pH is 7.

Strong acids, such as sulfuric acid, can burn the skin. Weak acids have many uses. For example, vinegar is weak acetic acid. Vitamin C is ascorbic acid. Citric acid makes orange, grapefruit, and lemon juices sour.

Weak bases include soap and shampoo. Bases are able to dissolve fats, which is why soap removes grease from dishes. Ammonia, which is often used in cleaning solutions, is a base. Sodium hydroxide, or lye, is a strong base. This base is used in drain cleaners.

Match the pH range of the substance with its description. Write the letter of the pH range in the blanks at the left. Answers may be used more than once.

		Description	pH range
______	26.	turns litmus paper red	a. acid
______	27.	turns litmus paper blue	b. base
______	28.	water	c. neutral substance
______	29.	ammonia	
______	30.	vitamin C	
______	31.	shampoo	

Circle the number of the best answer.

32. Which of the following lists of substances is in order from the lowest pH to the highest pH?

(1) vinegar, water, soap, lye

(2) water, vinegar, soap, lye

(3) water, soap, vinegar, lye

(4) soap, lye, vinegar, water

(5) vinegar, soap, water, lye

Go on to the next page.

Electric Power

What do you actually buy when you pay your electric bill? You are buying electric energy, which is measured in kilowatt-hours.

The kilowatt is a measure of electric power. You can see how much power a light bulb uses by looking at the number on it. Common light bulbs range from 50 to 300 watts. One kilowatt is equal to 1,000 watts. If you use ten 100-watt bulbs, you are using one kilowatt of power. If you use five 200-watt bulbs, you are also using one kilowatt of power.

The electric company bills you for how much power you use and how long you use it. This is where the kilowatt-hour comes into the picture. If you use one kilowatt of power for one hour, you have used one kilowatt-hour of electric energy. The longer you use an appliance, the more energy you use. The electric meter keeps track of the kilowatt hours. Your bill is based on the meter reading.

The rate per kilowatt-hour is shown on the electric bill. Some companies charge more for electric energy in the summer when many people run their air conditioners. You can check your bill by multiplying the rate per kilowatt-hour by the number of kilowatt-hours you used.

Fill in the blank with the word or words that best complete each statement.

33. The ________________ is a measure of electric power.

34. The ______________________________ is a measure of electric energy use.

Circle the number of the best answer.

35. What determines the amount you owe on your electric bill?

    (1) the amount of power you use

    (2) how long you use the power

    (3) the rate per kilowatt-hour

    (4) options (1) and (2)

    (5) options (1), (2), and (3)

Check your answers on pages 233–235.

POSTTEST
Correlation Chart

Science

The chart below will help you determine your strengths and weaknesses in the four content areas of science.

Directions

Circle the number of each item that you answered correctly on the Posttest. Count the number of items you answered correctly in each row. Write the amount in the Total Correct space in each row. (For example, in the Life Science row, write the number correct in the blank before *out of 15*). Complete this process for the remaining rows. Then add the 4 totals to get your Total Correct for the whole 35-item Posttest.

Content Areas	Items	Total Correct	Pages
Life Science (Pages 12–97)	1, 2, 3, 4, 5 6, 7 8, 9, 10 11, 12 13, 14, 15	_____ out of 15	Pages 14–19 Pages 44–49 Pages 62–67 Pages 68–73 Pages 80–85
Earth Science (Pages 98–133)	16, 17, 18, 19 20, 21, 22	_____ out of 7	Pages 112–117 Pages 124–129
Chemistry (Pages 134–169)	23, 24, 25 26, 27, 28, 29, 30 31, 32	_____ out of 10	Pages 142–147 Pages 154–159
Physics (Pages 170–199)	33, 34, 35	_____ out of 3	Pages 184–189
TOTAL CORRECT FOR POSTTEST _____ out of 35			

If you answered fewer than 32 items correctly, determine in which areas you need further practice. Go back and review the content in those areas. Page numbers for specific instruction in those areas of science are given in the right-hand column above.

ANSWERS AND EXPLANATIONS

INVENTORY

PAGE 1

1. nucleus
2. cell wall
3. **(3) vacuole is empty.** According to diagram, the vacuole stores water and minerals, so the plant would need more water if the vacuole were empty. Options 1, 2, 4, and 5 are incorrect because these structures have functions other than storing water.

PAGE 2

4. contracting
5. relaxing
6. **(2) at the front of the thigh** According to the diagram, the paired muscles in the arm are on opposites sides of the arm. From this, you can infer that muscles that do opposite jobs are in opposite positions. If the muscles at the back of the thigh bend the knee, muscles at the front of the thigh would straighten the knee. Option 1 is incorrect because these muscles bend the knee. Options 3 and 4 are incorrect because these muscles move the ankle joint. Option 5 is incorrect because muscles around the hip bend the hip joint.

PAGE 3

7. Lactic acid
8. bacteria
9. **(1) Pasteurizing kills bacteria with high temperatures.** According to the article, pasteurizing involves heating and then quickly cooling the milk. The heat kills the bacteria. The rapid cooling is needed so that the milk does not cook. Option 2 does not kill the bacteria. Options 3 and 4 are incorrect because the bacteria are killed, not slowed. Option 5 is incorrect because no chemicals are involved in pasteurization.

PAGE 4

10. insects
11. insects
12. wind
13. insects
14. **(5) anthers and ovary** According to the article, these structures make the sperm and egg, which are needed for reproduction. Options 1 and 4 are incorrect because petals are not necessary for reproduction. The article states that some flowers have no petals. Options 2 and 3 are incorrect because both parts are necessary for reproduction.

PAGE 5

15. forest
16. grassland
17. cactus
18. **(2) drier** The article states that the grassland is drier than the forest, and the desert is drier than the grassland. Options 1, 3, and 4 are incorrect because the article does not indicate a pattern in these conditions. Option 5 is incorrect because there is only one correct answer.

PAGE 6

19. c
20. e
21. b
22. a
23. d

PAGE 7

24. sedimentary rocks
25. metamorphic rocks
26. **(1) granite** Options 2 and 4 are incorrect because limestone and sandstone are sedimentary rocks. Option 3 is incorrect because marble is a metamorphic rock. Option 5 is incorrect because granite is an igneous rock.

27. **(4) volcanoes.** According to the article, volcanoes contain melted rock. When melted rock hardens, igneous rocks form. From this, you can infer that an area with many igneous rocks once had volcanoes. Options 1 and 2 are incorrect because rivers and sediment are associated with the formation of sedimentary rocks. Option 3 is incorrect because pressure is associated with the formation of metamorphic rocks. Option 5 is incorrect because earthquakes are not associated with the formation of rocks.

PAGE 8

28. b
29. c
30. a
31. c
32. a

PAGE 9

33. **(4) solid and liquid only** According to the table, these states of matter have a definite volume which does not change. Options 1 and 2 are incorrect because they list only one of the states with a definite volume. Options 3 and 5 are incorrect because the volume of a gas changes as the gas expands to fill its container.
34. **(1) Its shape changes from not definite to definite.** According to the table, a liquid does not have a definite shape, but a solid does. Option 2 is incorrect because it is the opposite of what happens. Option 3 is incorrect because both solid and liquid have definite volumes. Option 4 is incorrect because the change is to the lowest temperature range. Option 5 is incorrect because there is a change in a property.

PAGE 10

35. force
36. acceleration
37. **(1) acceleration** According to the article, when an unbalanced force acts on an object, the object's motion changes. This is acceleration. All other options are examples of forces, not effects of forces.

UNIT 1: LIFE SCIENCE

SECTION 1

PAGE 14

1–2. Answers should be things you knew before reading the article.
3. Observation
4. Hypothesis
5. Experiment
6. Conclusion
7. The photo shows an incubator aboard the *Discovery*.
8–9. Questions should be things you expected the article to answer.

PAGE 16

There are many possible answers. Sample answers: Vellinger figured out a way to test the effect of weightlessness on embryos. To test the effect of weightlessness, you need to go up into space. It is possible to have children in outer space.

PAGE 17

You may have circled many words. Here are the words in dark print that you should have circled, along with their definitions.
Scientific method: a logical way of getting information and testing ideas.
Observation: the gathering of information.
Hypothesis: a guess about the answer to a question.
Embryo: an organism in the early stages of development.
Experiment: a test of a hypothesis.
Experimental group: in an experiment, the group being tested.
Control group: in an experiment, a group similar to the experimental group except for one thing.
Conclusion: the results of an experiment.

PAGES 18–19

1. observation
2. hypothesis
3. experiment
4. conclusion

5. You should have written the answers to the questions you wrote on page 14 or three things you learned from reading the article.
6. In the future, people may need to spend long periods of time in space, either in flight or on space stations.
7. All of the eggs hatched except the younger group of embryos that had been weightless for five days.
8. **(2) solve problems in a logical way.** According to the article, the scientific method is a logical way of gathering information and testing ideas. Option 1 is incorrect because it is not mentioned. Options 3 and 5 are incorrect because they are only part of the scientific method. Option 4 is incorrect because the scientific method is a general procedure.
9. If the results in the two groups differ and there is only one difference between the two groups, you can be fairly sure that the difference in results was caused by the difference between the control and experimental groups.

10–11. There are many possible answers. Sample answers for 11: Yes, because it would be exciting. Yes, because I might become famous. No, because it is too dangerous.

SECTION 2

PAGE 20

1–2. Answers should be things you knew before reading the article.
3. How Sunlight Causes Skin Cancer
4. Preventing Skin Cancer
5. The diagram shows the parts of a cell.

6–8. Questions should be things you expected the article to answer.

PAGE 22

You should have underlined the following sentences:
Squamous cell skin cancer is more dangerous than basal cell skin cancer.
By far the most dangerous of the three types of skin cancer is melanoma.

PAGE 23

How Sunlight Causes Skin Cancer

PAGES 24–25

1. Ultraviolet light
2. Basal cell skin cancer
3. melanoma
4. cell
5. nucleus
6. Skin cancer
7. There are many possible answers.
8. Exposure to ultraviolet light can cause skin cancer.
9. You should have written one of the facts from the paragraph.
10. The cells divide abnormally.
11. The three types of skin cancer are basal cell, squamous cell, and melanoma.
12. **(4) going to a tanning parlor weekly** According to the article, people should not use tanning parlors. Options 1–3 and 5 are wrong because they are actions that help prevent skin cancer.
13. Other parts of the body may be covered, but the face is almost always exposed to the sun when a person is outdoors during the day.

14–15. There are many possible answers. Answers to 14 should be chosen from the list in the article.

SECTION 3

PAGE 26

1–2. Answers should be things you knew before reading the article.
3. Help with Reading the Labels
4. The Effect of Fat and Cholesterol
5. Food Companies Respond
6. What You Can Do
7. The photos show cross-sections of healthy arteries and arteries clogged with plaque.

8–9. The questions should be things you expected the article to answer.

PAGE 28

You should have listed three of the following pages: 1, 22–23, 28, 92

PAGE 29

Saturated fat makes up 19 percent of the calories in the regular hamburger.
The low-fat hamburger has more calories from carbohydrates.

PAGES 30–31

1. Saturated fat
2. Polyunsaturated fat
3. arteries
4. Plaque
5. You should have written the answers to the questions you wrote on page 26 or three things you learned from reading the article.
6. Saturated fat raises the body's blood cholesterol level.
7. The body makes cholesterol from saturated fat.
8. Extra cholesterol travels around the body in the blood. It is deposited on the inside walls of arteries and clogs them.
9. **(2) eating fewer eggs** Eggs contain cholesterol. Eating fewer eggs will help reduce the body's cholesterol level and reduce the risk of heart disease. Options 1, 3, and 5 are incorrect because the foods contain saturated fat, which raises the cholesterol level. Option 4 is incorrect because fried foods are high in fat.
10. **(3) an artery is blocked and not enough blood reaches the heart.** Lack of blood causes part of the heart muscle to die. Options 1 and 4 are true, but they do not cause heart attacks. Option 2 is the opposite of what happens. Option 5 is not true because saturated fats can be digested.
11. This is unusual because most foods high in saturated fat come from animal products. Avocados are fruit.

12–13. There are many possible answers. Sample answers for 13: I would be willing to eat fewer eggs and less butter. It would be hard to give up ice cream.

SECTION 4

PAGE 32

1–3. You should have written three things you knew before reading the article.

4. The Development of an Unborn Baby
5. The Effects of Drugs
6. Preventing Birth Defects
7. The diagram shows how substances pass through the placenta.

8–10. Questions should be things you expected the article to answer.

PAGE 34

The main idea is that substances pass back and forth through the placenta between the mother and the embryo.
You should have circled the following labels: Placenta, Embryo, Fluid, Baby's blood vessels, Mother's blood vessels, Food and oxygen, and Wastes and carbon dioxide.

PAGE 35

You should have listed two of the following: Nicotine makes the blood vessels in the placenta shrink. Less oxygen and fewer nutrients reach the baby. Smoking has been linked to miscarriages and stillbirths.

PAGES 36–37

1. zygote
2. embryo
3. placenta
4. You should have written the answers to the questions you wrote on page 32 or three things you learned from reading the article.
5. Thalidomide and DES helped focus research on the effects of drugs on the unborn baby.
6. The most serious damage is likely to happen while the developing baby is an embryo.
7. **(4) cocaine.** According to the article, cocaine babies may have damage to the brain or other organs. Options 1 and 2 are incorrect because aspirin and aspirin substitutes are possibly linked to minor problems. Option 3 is incorrect because cigarettes are linked to low birth weight, but there is no mention of brain damage. Option 5 is incorrect because only one of the options is correct.
8. **(5) do all of the above.** According to the last part of the article on page 35, options 1–4 are all steps a pregnant woman should take to help ensure a healthy baby.

9. **(1) all of the body organs have formed.** On page 34, the article states that the main body organs and systems form while the developing baby is an embryo. The article also states that during the third to ninth month the fetus grows. There is no mention of organs forming during this time. From this information, you can infer that the organs have formed before the last few months of pregnancy. Options 2–5 are all untrue and cannot be inferred from reading the article.

10. There are many possible answers.

11–12. There are many possible answers. Sample for 11: It is a good law because some women may not know that alcohol is dangerous and this would warn them. The law does not work because people do not read the fine print on a label. Sample answers for 12: I would tell her that she should not smoke because it is bad for her baby. I think she should do what she wants, so I would not say anything.

SECTION 5

PAGE 38

1–3. You should have written three things you knew before reading the article.

4. The Benefits of Walking
5. Improving Bones and Muscles
6. Easy on the Joints
7. The diagrams show how leg muscles work.

8–10. Questions should be things you expected the article to answer.

PAGE 40

The implied main idea is that walking strengthens muscles.

PAGE 41

Walking and running are similar because both burn about the same number of calories per mile.

PAGES 42–43

1. c
2. a
3. e
4. d
5. b
6. You should have written the answers to the questions you wrote on page 38 or three things you learned from reading the article.
7. Walking makes the heart beat faster and harder, which makes the heart stronger.
8. You should have included two of the following reasons: Walking is simple to do. Walking can be done almost anywhere. Walking is easy on the body.
9. **(5) all of the above** According to the article, walking can provide all the benefits listed in options 1–4.
10. **(1) One leg muscle contracts and the other relaxes.** According to the article, muscles work in pairs, doing the opposite action, to make bones move. Options 2 and 3 are incorrect because the muscles are doing the same, instead of opposite, actions. Options 4 and 5 are incorrect because ligaments connect bones at the joints. Ligaments do not move bones.
11. **(2) Pump your arms.** According to the article, pumping your arms builds upper body muscles. Options 1, 3, 4, and 5 are incorrect because they do not strengthen the upper body muscles.
12. Running and walking are similar in that they burn about the same number of calories per mile. They are different because running takes less time and can cause more injuries.
13. There are many possible answers.

SECTION 6

PAGE 44

1–2. You should have written two things you knew before reading the article.

3. Catching a Cold
4. Treating a Cold
5. The Flu
6. Pneumonia
7. The photo on page 45 shows a group of people playing cards.

8–10. Questions should be things you expected the article to answer.

PAGE 46

You should have circled *A Rhinovirus*.
You should have underlined *Protein covering* and *Genetic material*.
The diagram shows the structure of a rhinovirus.

PAGE 47

There are many possible answers. Sample answers: Ten percent of flu-related pneumonia is caused by viruses. At the hospital, pneumonia patients are given help with their breathing.

PAGES 48–49

1. virus
2. Influenza (or The flu)
3. Pneumonia
4. Bacteria
5. antibiotics
6. antibody
7. You should have written the answers to the questions you wrote on page 44 or three things you learned from reading the article.
8. The parts of a bacterial cell are the cell membrane, cell wall, cytoplasm, and genetic material.
9. **(3) pneumonia.** According to the article, bacteria cause pneumonia. Options 1, 2, 4, and 5 are incorrect because colds and the flu are caused by viruses, not by bacteria.
10. **(1) colds.** According to the article, half of all colds are caused by rhinoviruses. Option 2 is incorrect because the flu is caused by other viruses. Option 3 is incorrect because pneumonia is caused by other viruses or by bacteria. Options 4 and 5 are incorrect because these combinations include the flu and pneumonia, which are not caused by rhinoviruses.
11. **(1) antibiotics do not fight viruses.** According to the article, antibiotics fight bacteria. Option 2 is incorrect because antibiotics do fight bacteria. Option 3 is incorrect because it has nothing to do with the question. People do not have to go to the hospital to get antibiotics. Options 4 and 5 are incorrect because they are not true.
12. The symptoms of a cold include a sore throat, runny nose, congestion, and coughing. There is little fever. The symptoms of the flu are similar, except they come on more quickly. There may be a high fever within twelve hours.
13. There are many possible answers.

SECTION 7

PAGE 50

1–3. You should have written three things you knew before reading the article.
4. How Traits Are Passed from Parent to Child
5. Testing for Inherited Disorders
6. Genetic Screening of the Unborn
7. The diagram shows a Punnett square.
8–10. Questions should be things you expected the article to answer.

PAGE 52

The h stands for the recessive form of the trait.
No, the child will not develop Huntington's disease.

PAGE 53

You should have circled the words *believe* and *feel*.
There are many possible answers. You should have written your own opinion, stating how you feel about genetic screening.

PAGES 54–55

1. Huntington's disease
2. Traits
3. heredity
4. Genetics
5. dominant
6. recessive
7. Punnett square
8. You should have written the answers to the questions you wrote on page 50 or three things you learned from reading the article.
9. A capital letter represents a dominant trait.
10. **(2) amniocentesis** According to the article, this is a test performed during pregnancy. Options 1 and 5 are genetic disorders, not types of genetic screening. Option 3 is a type of diagram used in genetics. Option 4 refers to the inheritance of traits.

11. **(3) two out of four** Two of the four boxes contain the dominant trait (H) for Huntington's. Therefore, the odds that a child will inherit the disorder are two out of four. All other options are incorrect because they do not tell the correct odds.
12. A person at risk for Huntington's might not want to know definitely that he or she will become sick and die in middle age.

13–14. There are many possible answers.

SECTION 8

PAGE 56

1–3. You should have written three things you knew before reading the article.

4. Using Plant Medicines
5. Parts of a Plant
6. Searching for Medicinal Plants
7. The table shows medicines, the plants they come from, and their uses.

8–9. Questions should be things you expected the article to answer.

PAGE 58

You should have listed any words you did not know along with your guesses about their meanings.

PAGE 59

There are many possible answers. Sample answer: Scientists test plant extracts to find out if they work and how much is safe to use.

PAGES 60–61

1. e
2. a
3. b
4. f
5. c
6. d
7. You should have written the answers to the questions you wrote on page 56 or three things you learned from reading the article.
8. Plants work by affecting the chemistry of the body or by affecting the bacteria or viruses that cause diseases.
9. Reserpine and valepotriates both come from the roots of plants and are used as sedatives.
10. **(3) curare** According to the table on page 57, curare is used during surgery to relax muscles. Option 1 is incorrect because quinine is used to treat malaria. Option 2 is incorrect because menthol is used in pain relievers and decongestants. Option 4 is incorrect because digitalis is used to treat heart disorders. Option 5 is incorrect because taxol is used to treat ovarian cancer.
11. **(5) two-thirds** According to the article, two-thirds of the plant species are found in rain forests.
12. **(4) The effectiveness and safety of new drugs must be known.** According to the article, drugs undergo years of testing so scientists can decide whether they are effective and safe. Options 1, 2, 3, and 5 are all true, but they are not reasons for testing drugs.
13. There are many possible answers.

SECTION 9

PAGE 62

1–3. You should have written three things you knew before reading the article.

4. The Hunter and the Hunted
5. The Wolf's Niche
6. The Behavior of Wolves
7. People and Wolves
8. The diagram shows a food chain.

9–11. Questions should be things you expected the article to answer.

PAGE 64

There are many possible answers. Sample answers: Mice, rabbits, squirrels, chipmunks, horses, cows, sheep, and chickens are herbivores.
Lions, tigers, leopards, cats, coyotes, foxes, seals, and sea lions are carnivores.

PAGE 65

The higher-ranked wolf has its head and tail up. The lower-ranked wolf is crouching, with its tail and head down.

PAGES 66–67

1. packs
2. predators
3. prey
4. niche

5. food chain
6. herbivore
7. carnivore
8. You should have written the answers to the questions you wrote on page 62 or three things you learned from reading the article.
9. There are many possible answers. Sample answer: There are three wolves sniffing animal tracks with their heads lowered and other wolves behind them looking out.
10. **(1) sun** According to the article, the main source of energy in a food chain is the sun. Options 2, 4, and 5 are parts of the food chain that pass along energy. Option 3 is made by plants using the sun's energy.
11. **(1) predator of large animals.** According to the article, Options 2–5 are not true.
12. **(3) gazelles** Gazelles are the herbivores because they feed on plants. Options 1 and 2 are incorrect because these things do not eat, so they cannot be herbivores. Option 4 is incorrect because lions are carnivores. Option 5 is incorrect because Option 3 is correct.
13. A wolf's sharp teeth and claws help it grab and hold onto its prey. The teeth and claws are also useful in tearing meat into pieces small enough to eat.
14. There are many possible answers.

SECTION 10

PAGE 68

1–3. You should have written three things you knew before reading the article.
4. The Spread of the Gypsy Moth
5. The Life Cycle of the Gypsy Moth
6. Controlling the Gypsy Moth
7. The title of the diagram is *The Gypsy Moth's Life Cycle*.

8–10. Questions should be things you expected this article to answer.

PAGE 70

The four stages are egg, caterpillar, pupa, and adult. The next stage would be the egg.

PAGE 71

You should have underlined *Life Cycle of a Gypsy Moth*.
The timeline tells the order and length of the stages. It tells the month when each stage happens.

PAGES 72–73

1. egg
2. pupa
3. parasite
4. You should have written the answers to the questions you wrote on page 68 or three things you learned from reading the article.
5. According to the timeline, the egg stage lasts the longest. The egg stage lasts from August through the middle of May, almost ten months.
6. b
7. a
8. d
9. c
10. **(2) caterpillar** According to the article, gypsy moths eat leaves during the caterpillar stage. That's when trees are stripped and most of the damage is done. Options 1, 3, and 5 are incorrect because the gypsy moth doesn't eat leaves during these stages. Option 4 is not a stage of the gypsy moth's life cycle.
11. **(1) chemical sprays** According to the article, chemical sprays are poisonous and can damage other organisms that come into contact with them. Options 2–5 are incorrect because they are control methods designed to harm only the gypsy moth.
12. **(2) keep males from mating with females, who then lay eggs.** According to the article, adult males do not eat leaves. However, destroying them will prevent them from mating. This will decrease the number of eggs that are laid. In turn, this should reduce the damage caused by the next generation of moths. Options 1 and 5 are not supported by the article. Option 3 is incorrect because adult gypsy moths do not eat. Option 4 is not a fact from the article.

13. Egg masses can be scraped off and soaked in kerosene, ammonia, or bleach. To figure this out, look at the timeline on page 71 to see which stage occurs during February. Then read the description of the control method used to kill eggs.
14. There are many possible answers.

SECTION 11

PAGE 74

1–3. You should have written three things you knew before reading the article.
4. What We Throw Away
5. What Happens in a Landfill?
6. What Happens in a Compost Pile?
7. What You Can Do
8. The chart shows what makes up municipal solid waste.

9–11. Questions should be things you expected the article to answer.

PAGE 76

You should have circled *Municipal Solid Waste: What's in It, 1970 and 1986.*
The measurement is millions of tons.
The words describe types of garbage.
The black bars show 1970 figures.
Glass, metals, and food waste decreased from 1970 to 1986.

PAGE 77

Composting is being applied to large-scale commercial waste disposal in ten states.

PAGES 78–79

1. recycled
2. incinerated
3. landfills
4. Inorganic
5. organic
6. Decomposers
7. You should have written the answers to the questions you wrote on page 74 or three things you learned from reading the article.
8. Paper accounted for the most garbage in 1970.
9. A landfill is too deep for oxygen to reach the garbage. Oxygen is needed by aerobic bacteria to decompose garbage.
10. **(2) 13 percent** According to the article, 13 percent of garbage is recycled. The other options are incorrect.
11. **(4) plastics** Plastics went from about three million tons to ten million tons, a threefold increase. Options 1, 2, and 5 increased, but they did not triple. Option 3 decreased.
12. **(2) are turned to maintain oxygen levels.** Since aerobic bacteria need oxygen to degrade garbage, they work well in compost piles. In landfills there is not enough oxygen below a certain depth for aerobic bacteria to survive. Options 1, 3, and 4 are not true. Option 5 is true, but it does not answer the question.

13–14. There are many possible answers. Sample answer for 13: Yes, because recycling is very important, and I should do all I can. No, because I really like my brand, and I don't think it's worth it to change.

SECTION 12

PAGE 80

1–3. You should have written three things you knew before reading the article.
4. What Is a Tropical Rain Forest?
5. The Global Effects of the Rain Forests
6. Destruction of the Rain Forests
7. Stopping the Destruction
8. The diagram shows the carbon dioxide–oxygen cycle.

9–10. Questions should be things you expected the article to answer.

PAGE 82

More forests would decrease the amount of carbon dioxide in the air because the extra plants would absorb more during photosynthesis.

PAGE 83

You should have circled *The World's Tropical Rain Forests*. Tropical rain forests are shown in color. The rain forests are along or near the equator.

PAGES 84–85

1. ecosystem
2. Photosynthesis
3. Respiration
4. carbon dioxide–oxygen cycle
5. You should have written the answers to the questions you wrote on page 80 or three things you learned from reading the article.
6. Rice, bananas, coffee, and sugar first came from the rain forests.
7. Animals use oxygen and give off carbon dioxide during respiration. When they decay, they also give off carbon dioxide.
8. **(1) photosynthesis.** Photosynthesis absorbs carbon dioxide, thus decreasing the amount of carbon dioxide in the air. The other processes increase the amount of carbon dioxide in the air.
9. **(1) increased.** According to the article, burning fuel gives off carbon dioxide. So the increase in burning fuel has increased the amount of carbon dioxide in the air. The other options are incorrect.
10. **(3) All tropical rain forests are located south of the equator.** According to the map, there are tropical rain forests both north and south of the equator. All the other options are true.
11. There are many possible answers. Sample answers: Tropical rain forests have a wealth of plant and animal life. Tropical rain forests are the source of many medicines. Tropical rain forests are the source of many foods. Tropical rain forests help control the temperature of Earth.
12. There are many possible answers. Sample answers: I can buy rain forest products that are harvested without damaging the forest. I can refuse to buy furniture made from tropical woods.

SECTION 13

PAGE 86

1–2. You should have written two things you knew before reading the article.
3. The City Rat and the Country Rat
4. Rat Control
5. Adapting to Warfarin
6. Evolving Rats
7. Changing Tactics
8. The drawings show how rats evolve and develop resistance to Warfarin.

9–10. Questions should be things you expected the article to answer.

PAGE 88

There are many possible inferences. Sample answers: Something was wrong with the Warfarin. The Warfarin was no longer poisonous to these rats.

PAGE 89

Some insects were resistant to the insecticide. They survived and produced resistant offspring.

PAGES 90–91

1. adaptation
2. mutation
3. Natural selection
4. evolution
5. You should have written the answers to the questions you wrote on page 86 or three things you learned from reading the article.
6. Rats come out at night when people are asleep.
7. The rats had become resistant to Warfarin.
8. **(2) white fur** According to the article, white fur helps the fox blend in with its surroundings. As a result, the fox's enemies have trouble seeing it. Option 1 does not have any particular survival value. Options 3–5 would hurt the fox's chance for survival.
9. **(4) reproduce.** Reproduction is necessary if the fittest are to pass the adaptation to the next generation. Options 1, 2, 3, and 5 are actions that do not affect the species, just the individual.
10. **(2) Warfarin is no longer present in the environment.** According to the article, an adaptation is neither good nor bad when it has no effect on the individual's chances of survival. The other options are incorrect.
11. Favorable adaptations spread quickly because many generations occur in a short period of time.

12. Desert animals have trouble finding water. The juicy cactus would be completely eaten if it did not have spines for protection.
13. There are many possible answers.

UNIT 1 REVIEW

PAGE 92

1. DNA
2. mitosis
3. **(3) the two new cells will have all the information they need.** According to the article, DNA contains all the instructions the cell needs to live. This information is duplicated so that the cells produced by division will have all the instructions they need to live. Options 1 and 2 are incorrect because the parent cell has all its DNA and is finished growing by the time it divides. Option 4 is not true. Option 5 is incorrect because it doesn't matter to the cell if the chromosomes are easy to see.

PAGE 93

4. artery (The labels indicate that arteries carry blood to the lungs or the other parts of the body. Therefore, arteries are carrying blood away from the heart.)
5. left atrium
6. **(4) ventricle** The diagram shows that a ventricle has very thick walls when compared to an atrium, an artery, or a vein. Options 1–3 are incorrect because an artery, vein, or atrium has a thinner wall. Option 5 is incorrect because the walls are not all of the same thickness.

PAGE 94

7. antibodies
8. immune
9. **(3) identifying the virus that causes AIDS.** Scientists have to identify the virus before they can do anything about it. Option 1 is incorrect because vaccines do not contain white blood cells. Options 2 and 5 are incorrect because you cannot weaken or kill something until you have identified it. Option 4 is incorrect because a vaccine does not contain antibodies.

PAGE 95

10. dominant
11. recessive
12. genes
13. **(3) each parent has one dominant gene and one recessive gene.** From the description of the parents, you can conclude that each parent has at least one dominant gene, for right-handedness. Their left-handed child must have two recessive genes for left-handedness. Each parent passed on one of these recessive genes to the child. From this information, you can tell that each parent has one dominant gene and one recessive gene.

PAGE 96

14. b
15. c
16. a
17. nitrogen-fixing bacteria

PAGE 97

18. food, shelter
19. protection
20. scavengers
21. **(1) A bird eats ticks that are on the back of an ox.** The bird gets food from this relationship, and the ox gets rid of parasites. Options 2 and 3 are incorrect because the only living species involved is the ant. Option 4 is incorrect because the lions don't get anything from the wild dog. Option 5 is incorrect because the two animals in the relationship are the same species, and mutualism is a relationship between two different species.

UNIT 2: EARTH SCIENCE

SECTION 14

PAGE 100

1–3. You should have written three things you knew before reading the article.
4. Air Masses

5. Fronts
6. What Does a Weather Map Show?
7. Weather Forecasts
8. The map shows temperatures and precipitation for June 28.

9–10. Questions should be things you expected the article to answer.

PAGE 102

The symbol for a warm front is a line of half circles.
Fargo is closest to a warm front.
The temperatures are in the 60s and 70s in the cold air mass.
The temperatures are in the 80s, 90s, and 100s in the warm air mass.

PAGE 103

You should have circled the cold front that stretches from Detroit to Boston.
The weather in New York City on June 29 is clear and in the 90s.
There are several possible answers. But it is likely that the front will continue to move southeast and pass New York City. On June 30, the weather is likely to be cooler, with temperatures in the 70s.

PAGES 104–105

1. a, c
2. b, c
3. a, d
4. b, d
5. air mass
6. front
7. meteorologists
8. forecast
9. You should have written the answers to the questions you wrote on page 100 or three things you learned from reading the article.
10. The weather in the United States generally moves from west to east.
11. **(3) a stationary front** According to the two weather maps in the article, a stationary front is shown by alternating triangles and half circles pointing in opposite directions. The other options have other symbols, so they are incorrect.
12. **(4) New York** New York has hot, clear weather with temperatures in the 90s, which is good for the beach. The cities in options 1–3 have cooler temperatures in the 60s and 70s. Option 5, Fargo, is an inland city with rainy, cool weather.
13. **(1) occasional showers, high temperature in the 60s** Salt Lake City is in the middle of an air mass. In this air mass, temperatures are in the 60s and 70s and there are showers. Even if the cold front moves, the air mass is not likely to pass Salt Lake City by the next day.
14. The cold air mass in the west is probably a maritime polar air mass. It is cool, indicating polar. And it is moist (showers), indicating maritime origins.
15. Answers will depend on local weather conditions.

SECTION 15

PAGE 106

1–3. You should have written three things you knew before reading the article.

4. How Earth Is Warmed
5. Are People's Activities Increasing the Greenhouse Effect?
6. What Can Be Done About the Greenhouse Effect?
7. The diagram shows the greenhouse effect.

8–10. Questions should be things you expected the article to answer.

PAGE 108

Sunlight is light from the sun.

PAGE 109

Ideas 4 and 5 are both results, or effects, in the atmosphere that are caused by human activity. The problems listed as ideas 1, 2, and 3 are the causes of the effects listed as ideas 4 and 5.

PAGES 110–111

1. atmosphere
2. radiant
3. infrared radiation
4. greenhouse effect
5. You should have written the answers to the questions you wrote on page 106 or three things you learned from reading the article.
6. Radiant energy warms you when you sit out in the sun.
7. The greenhouse effect traps heat in the atmosphere the way a blanket traps heat.

8. The major greenhouse gases are carbon dioxide, ozone, chlorofluorocarbons (CFCs), methane, and nitrogen oxide.
9. **(5) all of the above.** Options 1–3 are fossil fuels, and burning them increases carbon dioxide in the atmosphere. Option 4 increases carbon dioxide by decreasing the number of trees that can absorb it during photosynthesis.
10. **(1) photosynthesis** During photosynthesis, plants absorb carbon dioxide from the air. Options 2, 4, and 5 all increase the amount of carbon dioxide in the air. Option 3 describes the way Earth is warmed.
11. **(3) The polar ice caps would melt.** Options 1, 4, and 5 are not true. Option 2 is true, but it is not a cause of changes in sea level.

12–13. There are many possible answers. Sample answers for 13: Reduce energy consumption in the home and car. Recycle as much as possible. Plant some trees.

SECTION 16

PAGE 112

1–3. You should have written three things you knew before reading the article.

4. Water as a Resource
5. Water Supplies
6. Conserving Water
7. The circle graph shows Earth's water resources.

8–10. Questions should be things you expected the article to answer.

PAGE 114

There are many possible answers. Sample answers: A large amount of the rainfall is not stored as part of the water supply. Some areas get more rain than others. Some areas have more people, so the need for water is greater.

PAGE 115

You should have circled the fourth through the seventh sentences. (Words that express opinions are *feel*, *think*, and *argue*.) There are many possible answers.

PAGES 116–117

1. irrigated
2. resource
3. groundwater
4. renewable resource
5. reservoir
6. aqueduct
7. You should have written the answers to the questions you wrote on page 112 or three things you learned from reading the article.
8. Most of the water in California is used by farmers to irrigate crops.
9. A drought has caused water shortages in California.
10. Fresh water is found in rivers, lakes, groundwater, the atmosphere, and frozen in ice at the poles.
11. **(3) Tripoli.** According to the article, both Tripoli and Los Angeles get about 14 inches of rain per year. The other options are incorrect because the article doesn't mention the amount of rainfall in those cities.
12. **(4) Some areas get more precipitation than they need, while others do not get enough.** According to the article, water shortages are caused by the uneven distribution of rainfall in the United States. Even though there may be plenty of water on average, there is not enough in specific areas. Option 1 is incorrect because the United States, as a whole, gets plenty of rain. Options 2 and 3 are not true. Option 5 is true, but it is not a cause of water shortages according to the article.
13. **(1) sunlight** A renewable resource is one whose supply can be refilled. Sunlight never runs out, so it is a renewable resource. Options 2–5 all occur in limited amounts. Once they are used up, they are gone. They are nonrenewable resources.

14–16. There are many possible answers. Sample answer for 14: Sources of water include wells, groundwater, reservoirs, rivers, and springs.

SECTION 17

PAGE 118

1–3. You should have written three things you knew before reading the article.
4. What Is a Crystal?
5. How Crystals Are Formed
6. The Importance of Minerals
7. The diagram shows the shapes of crystals.

8–10. Questions should be things you expected the article to answer.

PAGE 120

You should have circled *Cubic*, *Tetragonal*, *Hexagonal*, *Orthorhombic*, *Monoclinic*, and *Triclinic*.
The cubic shape looks most like a cube.
The hexagonal shape looks like a tube with six sides.
Gypsum is a mineral with monoclinic structure.

PAGE 121

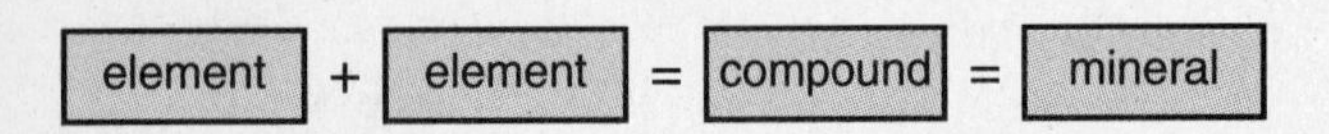

PAGES 122–123

1. mineral
2. atoms
3. compound
4. crystal
5. Ores
6. Metallic ores
7. Gems
8. You should have written the answers to the questions you wrote on page 118 or three things you learned from reading the article.
9. Crystals have a regular shape because their atoms are arranged in a pattern, not in a jumble.
10. **(3) hexagonal** According to the diagram on page 120, quartz is a mineral with the hexagonal shape. The diagram does not support the other options.
11. **(1) halite** According to the diagram on page 120, halite is a mineral with the cubic shape. Options 2–5 are incorrect since those minerals don't have the cubic shape.
12. **(4) options 1 and 2 only** According to the article, there are two ways that crystals are formed: (a) Crystals form when a liquid containing dissolved minerals evaporates. (b) Crystals form when liquid rock or metal cools and solidifies. Therefore, both options 1 and 2 are correct. Option 3 is incorrect because atoms in a crystal are arranged in a regular pattern.
13. There are many possible ways to diagram the relationships among ore, metallic ore, and nonmetallic ore. Sample diagram:

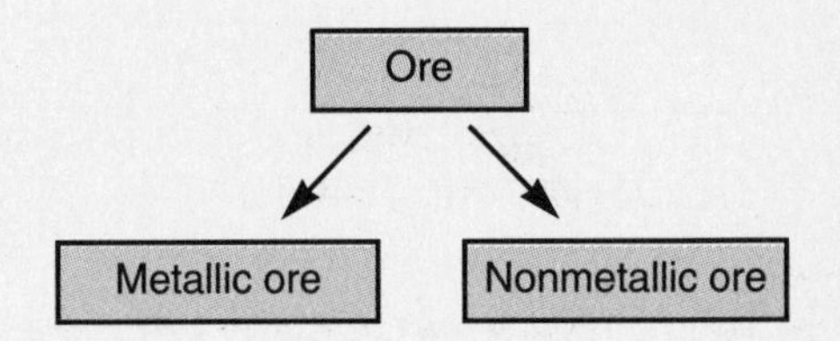

14. There are many possible answers. Sample answers: You might have such things as gold or silver jewelry, gems including birthstones, copper pennies, salt, pencils (graphite is the "lead" in a pencil), and so on.

SECTION 18

PAGE 124

1–3. You should have written three things you knew before reading the article.
4. The Mission
5. Jupiter and Saturn
6. Uranus and Neptune
7. The Mission Continues
8. The diagram shows where *Voyager 1* and *Voyager 2* went.

9–10. Questions should be things you expected the article to answer.

PAGE 126

You should have circled *Earth and Four Outer Planets.*
You should have drawn a box around Earth, Jupiter, Saturn, Uranus, and Neptune.
You should have underlined these characteristics: diameter, mass, distance from the sun, time of one revolution around the sun, time of one rotation, and number of known moons.
Jupiter has 16 moons.
Uranus takes 84 years to revolve around the sun.

PAGE 127

The farther the planet is from the sun, the longer it takes to revolve around the sun.

PAGES 128–129

1. c
2. b
3. e
4. a
5. d
6. You should have written the answers to the questions you wrote on page 124 or three things you learned from reading the article.
7. The solar system is made up of the sun and the planets and their moons.
8. The mission was to send photos and information about the planets Jupiter, Saturn, Uranus, and Neptune.
9. **(3) Uranus** According to the table, Saturn is 892 million miles from the sun. Uranus is 1,790 million miles from the sun. This distance is about twice as far from the sun as Saturn. Options 1, 2, and 4 are not supported by the table. Option 5 is incorrect because option 3 is the correct answer.
10. **(3) Saturn** According to the table, Saturn has 17 known moons, which is more than the other planets listed in the table. The other options are incorrect since the table shows they have fewer moons than Saturn.
11. **(4) operating a VCR or TV with a remote control** This is the most similar because you can turn the machine on and off, change channels, and change settings without touching the VCR or TV. Option 1 is not similar because a timer just starts a machine. You cannot change its operations with a timer. Option 2 requires you to touch the switch, and scientists could not touch either space probe in space. Option 3 simply places the camera. It does not operate the camera. Option 5 involves just receiving messages (music). It does not involve operating the machine providing the music. In addition, most headphones are attached by wires to the source of the music.
12. It is not practical because of the distances involved, the length of time required, and the hostile environments of the outer planets.
13. There are many possible answers.

UNIT 2 REVIEW

PAGE 130

1. snow
2. summer (Temperatures throughout the United States are very high. Temperatures are in the 90s as far north as Maine.)
3. **(2) hot, with showers** The map shows diagonal lines along the stationary front. According to the key, these lines indicate showers. The map also shows temperatures in the 80s, which is hot. Option 1 is incorrect because the map shows rain, not sunshine. Options 2 and 3 are incorrect because it is not cold along the front. Option 5 is incorrect because the map gives no indication of winds.

PAGE 131

4. oxygen
5. Ultraviolet
6. **(5) It is made up of CFCs.** According to the article, CFCs are damaging the ozone layer, but they do not form any part of it. Options 1–4 are incorrect because these are true statements which can be found in the article.

7. **(3) skin cancer.** According to the article, more ultraviolet rays reach Earth as CFCs destroy the ozone layer. These rays cause skin cancer. Options 1 and 2 are incorrect because poisoning is not part of the problem caused by CFCs. Option 4 is incorrect because ultraviolet rays do cause sunburn, but people do not die of sunburn. Option 5 is incorrect because only option 3 is correct.

PAGE 132

8. continental shelf
9. **(4) waters of the continental-shelf region.** According to the article, these waters have the richest fishing. Options 1 and 2 are incorrect because nodules, which are found only on the ocean basin, are mineral resources. They are not food resources. Options 3 and 5 are incorrect because these waters are not as rich a food resource as the waters of the continental shelf.
10. **(5) coal.** Options 1 and 5 are incorrect because these are resources from the continental shelf. Options 2 and 3 are incorrect because these minerals make up the nodules found on the ocean basin.

PAGE 133

11. sun
12. core
13. food
14. food chains
15. **(2) light** According to the article, most of the sun's energy is changed into light. A solar-powered calculator works only when placed in the light. Options 1 and 5 are incorrect because heat is not the form of energy that makes the calculator work. Option 3 is incorrect because nuclear reactions happen in the sun, but the sun does not cause them to happen on Earth. Option 4 is incorrect because this is not energy from the sun.

UNIT 3: CHEMISTRY

SECTION 19

PAGE 136

1–3. You should have written three things you knew before reading the article.
4. Substances and Mixtures
5. Chemical Symbols and Formulas
6. Chemical Reactions
7. The diagram shows different kinds of matter.

8–10. Questions should be things you expected the article to answer.

PAGE 138

Nitrogen and hydrogen are in ammonia. Ammonia has one nitrogen atom and three hydrogen atoms.

PAGE 139

Barium and chlorine are in barium chloride.
Calcium, chlorine, and oxygen are in calcium chlorate.
Carbon tetrachloride has four chlorine atoms.
A compound with one aluminum atom and three oxygen atoms is aluminum trioxide.

PAGES 140–141

1. Mixtures
2. substance
3. Elements
4. symbols
5. formula
6. reaction
7. equation
8. element
9. compound
10. compound
11. compound
12. You should have written the answers to the questions you wrote on page 136 or three things you learned from this article.
13. The symbol for sodium is Na.

14. **(5) lemonade** According to the article, a mixture contains different substances. Lemonade is a mixture of water, lemon juice, and sugar. Options 1–3 are compounds, and option 4 is an element.
15. **(2) magnesium sulfide** According to the article, when a compound is made of two elements, the suffix *-ide* is added to the root of the second element. Option 1 is incorrect because sulfur was not changed to sulfide. Option 3 is incorrect because there is only one atom of sulfur, not two. Options 4 and 5 are incorrect because sulfur is placed before magnesium, reversing the order in the formula.
16. **(5) iron and chlorine** The equation shows iron and chlorine combining to form iron chloride. The substances that react, or combine, are the reactants. Options 1 and 4 are incorrect because iron chloride is the product. Options 2 and 3 are only partially correct.
17. There are many possible answers. Sample answers: lye, detergent, bleach, ammonia, and baking soda. These cleaning compounds may be used to clean clothes, kitchens, bathrooms, floors, and many other household items.

SECTION 20

PAGE 142

1–3. You should have written three things you knew before reading the article.
4. Changes of State
5. Physical Changes and Chemical Changes
6. Cooking a Hamburger
7. The photos show melting ice and boiling water, which are two changes of state.

8–10. Questions should be things you expected the article to answer.

PAGE 144

In a chemical change, new substances are made.

PAGE 145

The browning of an apple is a chemical change. The dissolving of the sugar is a physical change.

PAGES 146–147

1. solid
2. liquid
3. gas
4. physical
5. chemical
6. evaporation
7. Freezing
8. physical change
9. physical change
10. chemical change
11. You should have written the answers to the questions you wrote on page 142 or three things you learned from reading the article.
12. **(4) Chemical and Physical Changes in Cooking** This title covers all the topics discussed in the article. Options 1–3 and 5 are too specific. They cover only parts of the article, so they are incorrect.
13. **(4) a physical change** The bubbles of gas in the soda are simply escaping from the soda. The make-up of the soda and the gas are not changing. Options 1 and 2 are incorrect because they involve changes of state. Matter is not changing state in this example. Option 3 is incorrect because the make-up of the soda and the gas remains unchanged. Option 5 is incorrect because oxygen is not being absorbed by the soda.
14. **(4) cooking an egg** According to the article, cooking an egg involves chemical changes in the egg's protein as it sets. Options 1–3 and 5 are all physical changes, so they are incorrect.
15. There are many possible answers. Sample answers: Slicing a loaf of bread, chopping parsley, freezing leftover stew, boiling water, melting chocolate, and popping a balloon are all examples of physical changes.
16. There are many possible answers.

SECTION 21

PAGE 148

1–3. You should have written three things you knew before reading the article.
4. What Is a Solution?
5. How Solutions Form

6. Solubility
7. The photos show different mixtures.

8–10. Questions should be things you expected the article to answer.

PAGE 150

Both mixtures and solutions are made up of two or more substances that can be separated by physical means. A solution is made of substances that are evenly distributed, and a mixture is made of substances in any proportion.

PAGE 151

You should have circled "Solubility of Some Solids in Water."
The labels along the side show solubility in ounces per quart of water.
The horizontal axis shows temperature in degrees Fahrenheit. About six ounces of sodium nitrate will dissolve.

PAGES 152–153

1. mixture
2. Distillation
3. solution
4. solvent
5. solute
6. solution
7. mixture
8. solution
9. You should have written the answers to the questions you wrote on page 148 or three things you learned from reading the article.
10. **(1) separating alcohol from brewed grain** According to the article, alcohol is separated from the fermented mixture by the process of distillation. Option 2 is incorrect because it is a chemical reaction. Options 3–5 are incorrect because they don't involve boiling a liquid, which is part of distillation.
11. **(3) zinc dissolved in copper** These two metals are solid at room temperature, so the solution is a solid under these conditions. Options 1, 2, 4, and 5 are not correct because the solvents are liquids or gases at room temperature.
12. **(4) 6 ounces** The answer to this question can be found by reading the solubility curve of sucrose (sugar) on the graph on page 151. The other options are incorrect.
13. Some Middle Eastern countries use distillation to purify their drinking water. Seawater could be distilled. Pure water would boil off, and then be condensed and collected. (Evaporation is also a possible answer, but it is too slow for the large-scale production of fresh water.)
14. There are many possible answers.

SECTION 22

PAGE 154

1–3. You should have written three things you knew before reading the article.

4. Combustion Reactions
5. How a Combustion Heater Works
6. Indoor Air Pollution
7. Safety Precautions
8. The photo shows an example of combustion.

9–10. Questions should be things you expected the article to answer.

PAGE 156

There are many possible answers. Sample answer: Since the article just discussed the advantages of kerosene heaters, it is likely that the article will discuss the disadvantages of kerosene heaters next.

PAGE 157

Complete combustion produces carbon dioxide and water vapor. Incomplete combustion produces other substances in addition to carbon dioxide and water vapor.

PAGES 158–159

1. combustion
2. hydrocarbon
3. activation
4. kindling
5. You should have written the answers to the questions you wrote on page 154 or three things you learned from reading the article.
6. Oxygen from the air is needed for combustion.

7. The hydrocarbon mixes with oxygen from the air. The hydrocarbon burns, producing carbon dioxide and water vapor. Heat and light are released.
8. For complete combustion, the fuel must be pure and there must be enough oxygen.
9. **(3) solar battery** The solar battery is the only device listed that does not burn a fuel to produce heat. All of the other options burn a fuel to give off heat, so they are incorrect.
10. **(1) carbon dioxide and water vapor** According to the article, these are the products of a complete combustion reaction. Option 2 is incorrect because sodium chloride is not a product of any kind of combustion. Options 3–5 are incorrect because they list products of incomplete combustion (carbon monoxide, nitrogen oxide, and sulfur dioxide) in place of carbon dioxide and water.
11. **(4) carbon monoxide** According to the article, carbon monoxide is one of the products of incomplete combustion in a kerosene heater. Options 1 and 5 are incorrect because they are outdoor pollutants. Options 2 and 3 are incorrect because they are not pollutants.
12. There are many possible answers. Sample answers: Heaters should be placed out of traffic and away from anything that might catch fire. Keep doors or a window open when using a combustion heater. Use high-quality kerosene in a portable heater.
13. There are many possible answers.

SECTION 23

PAGE 160

1–3. You should have written three things you knew before reading the article.
4. A Modern Energy Source
5. Nuclear Reactions
6. How a Nuclear Power Plant Works
7. The diagram shows a nuclear power plant.

8–10. You should have written three things you expected the article to answer.

PAGE 163

1. The fission reaction heats the coolant.
2. The coolant carries heat to a heat exchanger.
3. The heat from the coolant boils water.
4. Steam drives a turbine.
5. The turbine drives a generator that makes electricity.

PAGES 164–165

1. Nuclear energy
2. radioactive
3. Nuclear reactions
4. protons
5. Neutrons
6. fission
7. You should have written the answers to the questions you wrote on page 160 or three things you learned from reading the article.
8. One fission gives off enough neutrons to cause more fissions, so the reaction continues.
9. Plutonium is used as fuel in reactors.
10. **(5) d, c, a, b** According to the article and diagram, fission heats the coolant, heat from the coolant boils water, steam turns the turbine, and then the generator produces electricity. All of the other options are incorrect because the steps are out of sequence.
11. **(1) condense steam.** According to the diagram, river water is used in the condenser to cool the steam so it becomes water. Options 2 and 3 are incorrect because the water that does these things is not released. Options 4 and 5 are incorrect because nothing in the diagram indicates a cleaning process.
12. **(1) neutrons.** According to the article, control rods keep the reaction at a steady rate. The control rods absorb neutrons to keep the chain reaction from becoming too fast. Options 2–4 are incorrect because these materials are not absorbed by control rods. Option 5 is incorrect because option 1 is correct.
13. There are many possible answers.

UNIT 3 REVIEW

PAGE 166

1. mixture
2. graphite, clay
3. **(5) all of the above** According to the article, increasing the amount of clay makes the pencil harder. Harder pencils are given higher numbers. The lines made by harder pencils are finer and smudge less.

PAGE 167

4. exothermic reaction
5. endothermic reaction
6. chemical reaction
7. combustion
8. **(4) photosynthesis in a plant** According to the article, photosynthesis is an example of an endothermic reaction. All of the other options are incorrect because these reactions are exothermic and give off energy.

PAGE 168

9. **(5) oxygen, fuel, and heat.** According to the article, these are the three requirements for combustion. Options 1 and 2 are incorrect because carbon dioxide puts out fires instead of keeping them burning. Options 3 and 4 are incorrect because they are incomplete answers.
10. **(2) removing oxygen.** According to the article, small fires can be smothered. Throwing a heavy blanket on a fire would smother it. Option 1 is incorrect because the blanket does not add carbon dioxide. Options 3 and 4 are incorrect because the fuel and heat will be under the blanket. Option 5 is incorrect because the opposite is true.

PAGE 169

11. nuclear fusion
12. sun
13. helium

UNIT 4: PHYSICS

SECTION 24

PAGE 172

1–3. You should have written three things you knew before reading the article.

4. Force and Work
5. Simple Machines
6. A Compound Machine
7. The drawings show a lever and a wheel and axle.

8–10. You should have written three things you expected the article to answer.

PAGE 174

Some wood is scraped away, which sharpens the pencil.

PAGE 175

There are many possible answers. Sample answers: The smaller the front gear, the easier it is to pedal. The smaller the back gear, the harder it is to pedal. You need the lowest speed when the resistance is greatest.

PAGES 176–177

1. force
2. effort
3. resistance
4. simple machine
5. compound
6. mechanical advantage
7. gravity
8. gears
9. You should have written the answers to the questions you wrote on page 172 or three things you learned from reading the article.
10. The flush handle of a toilet is a lever.
11. A doorknob is a wheel and axle.
12. **(4) speed at one time and force at another time.** According to the article, a machine can multiply speed or force, but not both at once. Options 1 and 2 are incorrect because they are incomplete. Option 3 is incorrect because only one factor can be multiplied at a time. Option 5 is incorrect because option 3 is correct.
13. **(1) friction.** According to the article, friction is a force between surfaces that touch. Friction causes you to use extra effort. Options 2 and 3 are incorrect because these forces can act even if objects aren't touching. Option 4 is incorrect because effort is the force you exert to do work. Option 5 is incorrect because an advantage is not a force.

14. **(1) if an object moves.** According to the article, the definition of work says that an object moves. If the resistance is larger than the effort, there will be no motion. Options 2 and 3 are incorrect because effort and resistance are involved whenever work is done. Options 4 and 5 are incorrect because the forces of friction and gravity are not needed for work to be done.
15. Standing on the pedals adds to your effort force because you can use your weight—the downward pull of gravity—to move the pedal.
16. There are many possible answers.

SECTION 25

PAGE 178

1–3. You should have written three things you knew before reading the article.
4. Collisions
5. Corking Bats
6. Is Cheating Worth It?
7. The photo shows two players colliding.

8–10. Questions should be things you expected the article to answer.

PAGE 180

A Super Ball is very elastic. (This conclusion is based on the facts that elastic objects lose very little momentum in a collision and that a Super Ball loses little momentum as it collides with the ground.)

PAGE 181

There are many possible answers. Sample facts: A player was caught when his bat cracked. A corked bat is lighter than a solid bat. Sample opinions: One retired player believes his corked bat gave him extra home runs. Some scientists think that the lighter weight makes the corked bat better.

PAGES 182–183

1. collision
2. momentum
3. elastic
4. You should have written the answers to the questions you wrote on page 178 or three things you learned from reading the article.
5. A player corks a bat by cutting off the top and hollowing out the bat. Then the space is filled with cork and the top of the bat is replaced.
6. **(5) options (1) and (3)** An object's momentum is increased when either its speed or its weight is increased. Options 1 and 3 are only partly correct. Option 2 is incorrect because it is not true. Option 4 is incorrect because option 2 is incorrect.
7. **(4) and the other object may or may not be moving.** Options 1–3 are true of some collisions but not all collisions. Option 5 is incorrect because one or both objects may move after the collision.
8. **(3) Scientists have found that corking a bat makes it lighter than a solid bat.** This statement represents a measurement, which is a fact. The other options are statements of what people believe, think, or feel. These words indicate opinions.
9. The limousine has more momentum because it is a much heavier car.
10. If it were elastic, the body would have returned to its original shape instead of remaining dented.
11. There are many possible answers. Sample answers: One car pushes the other off its path. Parts of the cars are thrown into the air. One car spins the other one around.

SECTION 26

PAGE 184

1–3. You should have written three things you knew before reading the article.
4. What Is Electricity?
5. Volts and Amps
6. Too Much of a Good Thing?
7. The photo shows power lines.

8–10. Questions should be things you expected the article to answer.

PAGE 186

Appliances made in Europe can work at higher voltages.

PAGE 187

The heat melts the thin strip of metal in the fuse.

PAGES 188–189

1. electric current
2. circuit
3. Voltage
4. amperage
5. You should have written the answers to the questions you wrote on page 184 or three things you learned from reading the article.
6. Fuses and circuit breakers are safety devices.
7. **(3) plastic** According to the article, plastic is an example of an insulator. All other options are examples of metals. The article states that metals are good conductors.
8. **(5) all of the above** Electricity, sound, light, and heat are all kinds of energy.
9. **(2) It would take fewer appliances to blow the fuse.** If each appliance adds a few more amps to the circuit, it would take fewer appliances to add up to 10 amps than to 15 amps. Option 1 is incorrect because it would be unsafe to use more appliances. Option 3 is incorrect because the fuse would blow well before there was any danger of fire. Option 4 is incorrect because the fuse doesn't change the voltage of the circuit. Option 5 is incorrect because there is an effect.
10. The electrician removes the fuses or opens the circuits so electricity can flow. Then it is safe to work on the wires.

11–12. There are many possible answers.

SECTION 27

PAGE 190

1–3. You should have written three things you knew before reading the article.

4. Getting in Step
5. How a Laser Works
6. Lasers in Medicine
7. The diagram shows how a laser produces light.

8–10. Questions should be things you expected the article to answer.

PAGE 192

You should have circled *How a Laser Produces Light*.
You should have underlined *Completely reflecting mirror* and *Partially reflecting mirror*.
Light leaves through the partially reflecting mirror.

PAGE 193

A laser beam can pass through the eye and repair a blood vessel without harming the eye.

PAGES 194–195

1. Light
2. frequency
3. wavelength
4. You should have written the answers to the questions you wrote on page 190 or three things you learned from reading the article.
5. There are many possible answers. Sample answers: Laser light is stronger than white light. Laser light spreads out less than white light. Laser light is all in step and white light is not.
6. **(3) a light bulb** According to the diagram on page 192, options 1 and 2 are incorrect because these two mirrors are part of a laser. Option 4 is part of a laser because the article states that atoms give off light. Option 5 is part of the laser because the article states that there is a burst of light or electricity. So there must be a source of this energy.
7. **(2) get hotter.** According to the article, heat is caused by the laser beam. This seals a broken blood vessel. The other options are incorrect because the article never states that body tissues become colder, give off light or electricity, or grow larger.
8. **(1) The metal and the eye absorb different wavelengths of light.** According to the article, different lasers are used for different purposes. The article also states that each material absorbs a different wavelength of light. Options 2 and 3 are incorrect because the article does not discuss these differences. Options 4 and 5 are incorrect because they are not true. All laser light is made up of waves that are in step.

9. The pulsed laser does not leave scars or damage the gums.
10. Answers should include two of the following: Laser light can repair the eye without an operation. The narrow beam can be aimed at a tiny point in the eye. Laser light does not damage healthy parts of the eye.
11. There are many possible answers. Sample answers: No, because it has not been tested enough. Yes, because it will be quicker than a drill and may hurt less.

UNIT 4 REVIEW

PAGE 196

1. simple machine
2. inclined plane
3. Friction
4. **(4) pushing a wheeled 75-pound cart up a ramp three feet long** Options 1, 2, and 5 are incorrect because they involve lifting, which takes more force than using an inclined plane. Option 3 is incorrect because friction causes you to need more force to push a box than a wheeled cart.

PAGE 197

5. **(3) weight and speed** According to the article, momentum depends on both weight and speed. Options 1 and 2 are incorrect because direction does not affect momentum. Options 4 and 5 are incorrect because how far the object has moved does not affect momentum.
6. **(1) Its momentum increases because the speed increases.** As the rock rolls down a hill, it moves faster. The faster it moves, the more momentum it has. Option 2 is incorrect because momentum does not decrease. Options 3 and 4 are incorrect because the rock does not get heavier. Option 5 is incorrect because the momentum does change.

PAGE 198

7. Sound
8. amplitude
9. **(2) high amplitude and short wavelength.** The amplitude determines the loudness of the sound. The higher the amplitude, the louder the sound. The wavelength determines if the sound is high or low in pitch. The shorter the wavelength, the higher the pitch.

PAGE 199

10. electromagnetic field
11. magnet
12. **(5) a compact disc** According to the article, a compact disc does not have a magnetic pattern, so it will not be damaged by the EM field of the metal detector. All other options have magnetic patterns and will be damaged.

POSTTEST

PAGE 200

1. observations
2. problem
3. experiment
4. data
5. conclusion

PAGE 201

6. **(2) the host cell has the raw materials for making viruses.** According to the article, the virus does not have raw materials of its own, so it depends on the host cell for these things. Options 1 and 3 are incorrect because the virus has the genetic material but not any raw materials. Options 4 and 5 are incorrect. The genetic material of the host cell cannot make viruses because it is different from the genetic material of the virus.
7. **(4) dies.** The article states that the cell breaks open. The diagram shows that the cell is destroyed when this happens. Therefore, the cell dies. Option 1 is incorrect because cells do not turn into viruses. Options 2, 3, and 5 are incorrect because the cell is too badly damaged when it breaks open to return to its original form and produce anything more.

PAGE 202

8. predator
9. prey
10. **(1) there would be more hares.** If many predators are removed, it becomes easier for the prey to survive, so there would be more hares. Option 2 is incorrect because it is the opposite of what would happen. Option 3 is incorrect because the loss of wolves would affect the hares first. If there were more hares, there would be more, not fewer, lynxes. Options 4 and 5 are incorrect because a disease that affects wolves might not affect the hares or the lynxes.

PAGE 203

11. incomplete metamorphosis
12. **(5) options 1 and 3** The article states that only the adult has wings and reproduces. Options 1 and 3 are both true, which is stated in option 5. Options 2 and 4 are incorrect because both nymph and adult are able to eat.

PAGE 204

13. **(5) Photosynthesis is the reverse of respiration.** According to the article, the substances produced by photosynthesis are used in respiration. The substances produced in respiration are used in photosynthesis. Option 1 is incorrect because these reactions are not the same. Option 2 is incorrect because only respiration needs oxygen. Option 3 is incorrect because only photosynthesis needs sunlight. Option 4 is incorrect because only respiration takes place in animals.
14. **(3) food.** Animals get both food and oxygen from plants, but only food is taken in through eating. Option 1 is incorrect because animals do not get carbon dioxide from plants. Plants get carbon dioxide from animals. Option 2 is incorrect because animals get oxygen by breathing, not by eating. Option 4 is incorrect because sunlight can come only from the sun. Option 5 is incorrect because only one option is correct.
15. **(4) oxygen.** According to the article, in respiration, plants use oxygen and sugar. Options 1–3 are incorrect because carbon dioxide, water, and sunlight are used in photosynthesis. Option 5 is incorrect because only one option is correct.

PAGE 205

16. Metallic ores
17. oxide
18. resource
19. **(3) to combine with oxygen from the ore.** According to the article, one of the uses of coke is to combine with oxygen from the ore. Options 1, 2, and 4 are incorrect because refining does not add anything to the ore or make anything combine with the ore. Option 5 is incorrect because option 3 is correct.

PAGE 206

20. solar eclipse
21. partial eclipse
22. moon, sun

PAGE 207

23. oxidation
24. **(3) The plastic wrap keeps the metal away from oxygen.** According to the article, oxidation takes place when a metal reacts with oxygen. Keeping these substances apart keeps the reaction from taking place. Options 1 and 2 are incorrect because keeping the metal warm or dry does not stop oxidation. Options 4 and 5 are incorrect because rust and copper do not have anything to do with the tarnishing of silver.
25. **(5) silver chloride** This is the only compound named that does not contain the word *oxide*. All other options are oxides, compounds formed by oxidation.

PAGE 208

26. a
27. b
28. c
29. b
30. a
31. b

32. **(1) vinegar, water, soap, lye** Of the four substances listed, vinegar is the only acid, so it must be listed first. Water is the only neutral substance, so it must be listed second. The two remaining substances are bases. Soap, which is the weaker base, is listed before lye, the stronger base. The other options are incorrect because they are not in the correct order.

PAGE 209

33. kilowatt
34. kilowatt-hour
35. **(5) options 1, 2, and 3** According to the article, the three factors that determine the size of the electric bill are the amount of power you use, how long you use it, and the rate per kilowatt-hour. Although the other options are correct, option 5 indicates that all three are correct.

Acknowledgments *(continued from page ii)*

p. 69 © Peter Arnold
p. 75 © John Griffin/The Imageworks
p. 81 Carl Frank/Photo Researchers
p. 87 Grant Heilman
p. 88 Irene Vandermolen/Photo Researchers
p. 98 NASA
p. 99 © Lawrence Migdale/Stock, Boston
p. 108L © Lisa Law/The Imageworks
p. 108R © Tom McHugh/Photo Researchers
p. 113 Tom McHugh/Photo Researchers
p. 119L British Information Service
p. 119R Smithsonian Institution
p. 120 Dept. of Library Services, American Museum of Natural History
pp. 125–127 NASA
p. 134 © 1983 Melissa Hayes English/Photo Researchers
p. 135 © Kevin Walsh/UC San Diego School of Medicine
p. 143 © 1991 T. Alexander
p. 145 Irene Bayer/Monkmeyer Press
p. 149L George Jensen/Royal Copenhagen
p. 149R © 1991 T. Alexander
p. 155 © Alan Pitcairn/Grant Heilman
p. 161 © Tim Davis/Photo Researchers
p. 170 © Globus Brothers/Stock Market
p. 171 NASA
p. 173 © Lionel Delevingne/Stock, Boston
p. 175 Courtesy, Carbondale Corporation
p. 179 Bettmann Archive
p. 181 Marc S. Levine/New York Mets
p. 185 Kenneth Murray/Photo Researchers
p. 187 © 1991 T. Alexander

Glossary

A

activation energy the energy necessary to produce a combustion reaction

adaptation a trait that makes a plant or an animal better able to live in its environment

aerobic something that needs oxygen to live

air mass a large body of air with similar temperature and moisture

amniocentesis a test performed on pregnant women that detects certain birth disorders

amp a unit of electric current

amperage the strength of a current of electricity expressed in amperes

amplitude the height of a wave

antibiotic a drug that fights bacteria

antibody a protein made by white blood cells that attack and kill invading germs

antigen a foreign protein, or germ

aqueduct a pipe that carries water from a reservoir

artery a large blood vessel that carries blood to all parts of the body

atmosphere the air around us

atom the smallest particle of an element

B

bacteria one-celled organisms

ball-and-socket joint a joint that allows movement in almost all directions; (example: hip)

bar graph an illustration that is used to compare sets of information

basal cell skin cancer a type of slow-growing cancer that often appears on the hands or face as an open sore, reddish patch, mole, or scar

biome a large region with a certain climate and certain living things

boiling the rapid change of matter from a liquid to a gas

C

carbon dioxide–oxygen cycle a process in which plants use carbon dioxide given off by other living things and make oxygen which is in turn used by the other living things

carnivore an animal that eats other animals

cause tells why something happened

cell the smallest unit of a living thing that can carry on life processes

cell membrane a layer around the cell that controls what may enter or leave the cell

cell wall the stiff outer layer around a plant cell that provides support for the plant

chain reaction a reaction that keeps itself going

chemical change a change in the property of matter. A chemical change makes new substances.

chemical equation a statement that shows the reactants and products of a reaction

chemical formula a group of symbols used to name a compound; (example: H_2O is the chemical formula for water)

chemical reaction a process in which elements or compounds are changed into other substances

chemical symbol a kind of shorthand chemists use in which one or two letters stand for an element

chemistry the study of matter and its changes

chloroplast a part of a plant cell that uses chlorophyll and sunlight to make food in the cell

cholesterol a fatlike substance found in all animals

circle graph a graph used to compare amounts; also known as a pie chart

circuit a path through which electricity can travel

classifying grouping things that are similar to help understand how they work

climate the usual weather of a region over a long period of time

cold front the leading edge of a cold air mass

collision the result of a moving object striking another object

combustion the chemical change also known as burning, in which oxygen reacts with fuel to create light and heat

compare to tell how things are alike

composting the breaking down of organic material by aerobic bacteria, fungi, insects, and worms

compound two or more elements combined chemically

compound machine a machine made up of several simple machines

conclusion a logical judgment made from facts

condensation the change from a gas to a liquid

conductors materials that carry electric current

conserve to use something wisely

context figuring out the meaning of an unknown word by using the rest of the words in a sentence

continental polar air mass a cold, dry air mass that forms over Canada and the northern United States

continental shelf the nearly flat area of the ocean bottom where the ocean meets the continent

continental slope the sloping area that extends from the continental shelf to the ocean basin

continental tropical air mass a warm, dry air mass that forms over the southwestern United States

contrast to tell how things are different

control group in an experiment, a group similar to the experimental group except for one thing

core the center of something such as the sun or Earth

crystal a solid with a regular shape and flat sides

cytoplasm a jellylike material that makes up most of a cell

D

decomposer a living thing that breaks down organic material

degrade a process in which organic material breaks down

details small pieces of information that explain or support the main idea

diagram a drawing that often shows steps in a process with arrows showing how one step leads to another

distillation the separation of liquid mixtures

DNA the genetic material found in chromosomes

dominant trait a trait that will appear in offspring if contributed by a parent

Down syndrome a disorder caused by an extra chromosome. Children born with Down syndrome are mildly to severely mentally retarded and may also have other health problems

E

Earth science the study of Earth and its surroundings

ecosystem an area in which living and nonliving things interact

effect tells what happened as a result of the cause

effort the force that is being used to do work

elastic a thing that can be stretched or compressed and will then return to its original shape

electric current electricity that flows through a wire or other object

electricity a kind of energy

electromagnetic field (EM field) the energy field around and caused by an electric current

element a substance that cannot be broken down into other substances by ordinary means, such as heating or crushing

embryo an organism in the early stages of development; a developing baby from the third to eighth week in the womb

EM field *see* electromagnetic field

endangered species a plant or an animal that is in danger of becoming extinct

endothermic reaction a process in which energy is taken in; (example: photosynthesis)

environment all of the living and nonliving things that make up a place

evaporation the slow change of a liquid to a gas

evolution the gradual change in a species over time

exothermic reaction a process in which energy is produced; (example: combustion)

experiment a procedure used to test a hypothesis

experimental group in an experiment, the group being tested

F

fact a statement about something that actually happened or actually exists

fat a substance that provides energy and building material for the body

fetus a developing baby from the third to ninth month

fission the splitting of an atom's nucleus

fixed joint a joint that does not allow the bones to move; (example: the seams in the skull)

flu *see* influenza

food chain the transfer of energy from one organism to another in the form of food

force a push or a pull

forecast a prediction, as of the weather

freezing the change in matter from a liquid to a solid

frequency the number of waves that pass a point in a certain amount of time

friction a force between surfaces that touch

front the leading edge of a moving air mass

fuel a source of energy

fuse a device that prevents electrical overloads

gas a state of matter that does not have a definite size or shape and expands to fill its container

gear a wheel with teeth; each gear turns on its own center

gem a colorful mineral that is often cut, polished, and made into jewelry; (examples: diamonds, sapphires, rubies)

gene a part of the genetic material that determines a particular trait

genetic screening tests that can tell if certain disorders are likely to be inherited

genetics the study of how traits are inherited

glossary an alphabetical listing of important words and their definitions located at the end of a text

graph a special kind of drawing that is used to compare information

gravity the pull of Earth

greenhouse effect the warming of Earth caused by the absorption of infrared radiation into gases in the atmosphere

groundwater water that is found underground; (examples: springs, wells, and aquifers)

herbivore an animal that eats plants

herd a large group of animals

hereditary capable of being passed from a parent to an offspring

heredity the passing of traits from parents to their young

hinge joint a joint that allows movement in only one direction; (example: knee)

Huntington's disease a disorder that slowly attacks the brain and causes loss of physical and mental functions

hydrocarbon a compound made only of hydrogen and carbon

hypothesis a guess about the answer to a question based on many observations

I

implied not stated

incinerate to burn, as in burning garbage

inclined plane a simple machine with a long, sloping surface that helps raise an object; (example: ramp)

inference the use of information to figure out things that are not actually stated

influenza (flu) an illness caused by a virus

infrared radiation the energy Earth radiates back into the atmosphere

inherit to manifest a trait or disease that was passed on from parent to child; (examples: hair color, Huntington's disease)

inner planets the planets between Earth and the sun; Mercury and Venus

inorganic material garbage from things that were never alive

insulator a material that does not carry an electric current

ion a charged particle formed from an atom

irrigate to bring water to an area of land

J

joint the place where two or more bones come together

K

kindling temperature the temperature at which a substance will burn

L

landfill an area of open land that is dug up and filled with layers of garbage

laser a tool that produces a narrow, strong beam of light in which all the waves have the same frequency and wavelength

lever a bar that turns on a pivot

life cycle the series of changes an animal goes through in its life

life science the study of living things and how they affect one another

ligament a strong band that connects bones at joints

light a form of energy that travels in waves and makes vision possible

line graph a graph using lines to show how something increases or decreases over time

liquid a state of matter that takes up a definite amount of space but does not have a definite shape. A liquid flows and takes the shape of its container.

M

main idea the topic of a paragraph

maritime polar air mass a cold, moist air mass that forms over the northern Atlantic Ocean and northern Pacific Ocean

maritime tropical air mass a warm, moist air mass that forms over the Caribbean Sea, the middle of the Atlantic Ocean, or the middle of the Pacific Ocean

mechanical advantage the number of times a machine multiplies your effort to do work

melanoma a fast-growing skin cancer that may appear as oddly shaped blotches

melting the change in matter from a solid to a liquid

metallic ore an ore that contains metal; (examples: silver, gold, iron, copper)

mineral a solid substance that has a specific composition and structure

mitochondria the parts of a cell that give the cell the energy it needs to grow and reproduce

mitosis the process by which a cell divides and therefore reproduces

mixture a combination of two or more kinds of matter than can be separated by physical means

molecule the smallest particle of a compound

momentum the property of a moving object that determines how long it will take to come to a stop when under the action of a force

monounsaturated fat a type of fat found in some vegetable products

municipal solid waste organic and inorganic materials that are thrown away as garbage

mutation a change in a gene

mutualism a relationship in which two species help each other

N

natural selection the survival of organisms best suited to their environment

neutron a particle that has no charge, found in the nucleus of an atom

niche the role an animal plays in its environment

nitrate a substance made by soil-dwelling bacteria using nitrogen from the air

nonmetallic ore an ore that does not contain metal; (example: sulfur)

nuclear energy energy that comes from inside an atom

nuclear fusion the reaction caused by two nuclei combining to make the nucleus of a larger atom

nuclear reaction a change in the nucleus, or center, of an atom

nucleus the control center of a cell. The nucleus contains genetic material.

observation the act of watching and reading to gather information and learn about something

ocean basin the bottom of the sea

opinion a statement that expresses what a person or group of people think or believe about a fact

ore a rock containing minerals that can be mined for profit

organic material garbage that comes from things that were once alive

osteoporosis a condition of brittle bones common to older people, especially women

outer planets the planets farther from the sun than Earth is; Mars, Jupiter, Saturn, Uranus, Neptune, and Pluto

oxidation the process in which a substance reacts with oxygen, causing the formation of a new compound called an oxide

ozone a form of oxygen

P

pack a group of animals that lives together in its own territory

parasite an organism that lives on or in another organism and harms it

photosynthesis the process by which plants use carbon dioxide, water, and energy from sunlight to make food

pH scale a measurement from 0 to 14 of the strength of an acid or a base

physical change a change in the appearance of matter without a change in the property of the matter; (example: the dissolving of sugar in water)

physics the study of energy and how it affects matter

pivot the point around which an object turns

pivot joint a joint that allows rotation from side to side; (example: neck)

placenta a structure that attaches the embryo to the uterus and allows substances to pass between the embryo and the mother

plaque deposits of cholesterol on the inside walls of arteries

pneumonia an infection of the lungs caused by viruses or bacteria

polyunsaturated fat a type of fat found in some vegetable foods and fish

precipitation water falling from the atmosphere in the form of rain, snow, or sleet

predator an animal that hunts other animals for food

prediction a guess about what may happen

previewing to look over text quickly, without reading it

prey animals other animals hunt to eat

product a substance that forms in a chemical reaction

proton a positively charged particle in the nucleus of atoms

Punnett square a diagram used to show all possible combinations of traits among children of two parents

R

radiant energy energy from the sun

radioactive the state of atoms that are in the process of decay, in which the atoms are giving off particles and energy in the form of radiation

radioactive decay particles given off by an unstable nucleus in an atom (an unstable nucleus has an unbalanced number of protons and neutrons)

reactant a substance that reacts in a chemical reaction

recessive trait a trait that will not appear if combined with a dominant trait

recycle to process materials to use again

refining the process of separating metal from oxygen and other substances in rocks

renewable resource a resource that does not get used up; (example: water)

reservoir a lake created by a dam

resistance a force that must be overcome to do work

resource a material that people need from Earth

respiration the process by which living things take in oxygen and release carbon dioxide to obtain energy

rhinovirus a virus that cause certain types of colds

ribosome a part of the cell that makes proteins the cell needs to grow

S

saturated fat a type of fat that is solid at room temperature

scientific method an organized way of solving problems; the process scientists use for getting information and testing ideas

sequence the order in which things happen

simple machine a device to do work; (example: lever)

sliding joint a joint that allows the small bones to slide over each other; (example: ankle)

solar system the sun and the objects that revolve around it, such as the planets and their moons

solid a state of matter that has a definite shape and takes up a definite amount of space

solubility the amount of a solute that will dissolve in a given amount of solvent at a given temperature

solute the substance in a solution that is present in the smaller amount

solution a type of mixture in which the ingredients are distributed evenly throughout

solvent the substance that is present in the greater amount in a solution

sound a sensation perceived by hearing; caused by vibrations

sprain a joint injury in which the ligaments are stretched or torn

squamous cell skin cancer a type of cancer that looks like raised, pink spots or growths that may be open in the center

stationary front the zone between two air masses, caused when the masses stop moving

substance matter that is of one particular type

summarize to condense or shorten a larger amount of information into a few sentences

supporting statements details that lead to a conclusion

T

table a type of graph that organizes information

timeline an illustration that shows when a series of events took place and the order in which they occur

topic sentence the sentence that contains the main idea in a paragraph

trait an inherited characteristic such as hair color and blood type

U

ultraviolet light a type of light in sunshine

uterus a woman's womb

V

vaccination an injected dose of dead or weakened disease-causing agent. The body reacts to a vaccination by forming antibodies to fight the disease.

vaccine a substance containing weakened bacteria or viruses

vacuole a part of a plant cell that stores water and minerals

virus a tiny particle of genetic material with a protein covering

voltage the amount of electric energy available to a whole circuit

W

warm front the leading edge of a warm air mass

wavelength the distance from the top of one wave to the top of the next wave

weather map a map showing where cold, warm, and stationary fronts are, as well as areas of high and low pressure

wheel and axle two objects that turn on the same center

work the process of using force to cause an object to move

Z

zygote a fertilized egg resulting when the sperm from the father joins with the egg produced by the mother

Index

D

F

G

H

I

J

K

L

M

N

U

V

W

Z